Sustainable Energy in Developing Countries

Environment

MIND

Society Economy

Munasinghe Institute for Development

'making development more sustainable'

Munasinghe Institute for Development (MIND) Series on Growth and Sustainable Development

Series Editor:

Mohan Munasinghe
Chairman, Munasinghe Institute for Development, Sri Lanka
Honorary Energy Advisor to the Government, Sri Lanka
Visiting Professor, Yale University, USA
Vice Chair, Intergovernmental Panel on Climate Change, Switzerland

This series of three volumes deals with the interaction between conventional economic growth and the more recent paradigm of sustainable development. A distinguished group of international experts drawn from a broad range of disciplines examine this nexus, and suggest ways for making development more sustainable in the future. The volumes cover both theory and practical applications at the global, national and local levels.

1. The Sustainability of Long Term Growth: socioeconomic and ecological perspectives – Mohan Munasinghe, Osvaldo Sunkel and Carlos de Miguel
(Edward Elgar, 2001)

2. Sustainable Energy in Developing Countries: policy analysis and case studies – Peter Meir and Mohan Munasinghe
(Edward Elgar, 2005)

3. Macroeconomic Policies for Sustainable Growth: analytical framework and case studies of Brazil and Chile – Mohan Munasinghe, Raul O'Ryan, Ronaldo Seroa da Motta, Carlos de Miguel, Carlos Young, Sebastian Miller and Claudio Ferraz
(Edward Elgar, 2005)

Sustainable Energy in Developing Countries

Policy Analysis and Case Studies

---

by

Peter Meier
Chief Economist, IDEA, Switzerland

Mohan Munasinghe
Chairman, Munasinghe Institute for Development (MIND)
Chief Energy Advisor to the Government of Sri Lanka, Sri Lanka

Edward Elgar
Cheltenham, UK • Northampton, MA, USA

Research study supported by the World Bank, Washington DC, and the Munasinghe Institute for Development (MIND), Colombo.

Published by
Edward Elgar Publishing Limited
Glensanda House
Montpellier Parade
Cheltenham
Glos GL50 1UA
UK

Edward Elgar Publishing, Inc.
136 West Street
Suite 202
Northampton
Massachusetts 01060
USA

A catalogue record for this book
is available from the British Library

Library of Congress Cataloguing in Publication Data

Meier, Peter, 1942–
Sustainable energy in developing countries : policy analysis and case studies / by Peter Meier and Mohan Munasinghe.
p. cm.
1. Energy policy–Sri Lanka. 2. Power resources–Environmental aspects–Sri Lanka. 3. Energy conservation–Sri Lanka. I. Munasinghe, Mohan, 1945– II. Title.
HD9502.S722M45 2004
333.79'095493–dc22

2004051018

ISBN 1 84376 753 8

Printed and bound in Great Britain by MPG Books Ltd, Bodmin, Cornwall

Contents

Figures

Tables

Boxes

Acknowledgements

This volume is based on studies partially funded by the World Bank and the Munasinghe Institute for Development (MIND). The authors express their thanks to the members of the study team:

A. Abeywardena, Ceylon Petroleum Corporation
R. Chandrasekera, Ceylon Petroleum Corporation
S. Chandrasiri, University of Colombo
Chitrupa Fernando, University of Oklahoma, USA
Shavi Fernando, Ceylon Electricity Board
D. Jayaweera, Transportation Policy Studies Centre
A. Kumarage, University of Moratuwa
R. M. Munasinghe, University of Cambridge, UK
J. Nanthakumar, Ceylon Electricity Board

Their valuable contributions to this volume are gratefully acknowledged in the individual chapters.

The authors also wish to thank R. P. Samarakkody of the National Building Research Organisation (NBRO) for his assistance with air quality data.

Finally, we are indebted to Nishanthi De Silva and Yvani Deraniyagala of MIND for helping to prepare the manuscript.

Abbreviations and acronyms

2S	two-stoke motorcycle
4S	four-stroke motorcycle
BoI	Board of Investment
CBA	cost–benefit analysis
CCCT	combined cycle combustion turbine
CBD	central business district
CCPI	Colombo consumer price index
CEB	Ceylon Electricity Board
CGPI	Greater Colombo price index
CKE	Colombo-Katanayake Expressway
CPC	Ceylon Petroleum Corporation
crore	10 million (South Asia)
CUTS	Colombo Urban Transportation Study
DMU	diesel multiple unit
DSD	Secretariat Divisions
DSM	demand-side management
ECNZ	Electricity Corporation of New Zealand
EER	economic rate of return
EMU	electric multiple unit
ESP	electrostatic precipitator
FGD	flue gas desulphurisation
GCMR	Greater Colombo Metropolitan Region
GDP	gross domestic product
GDPD	implicit gross domestic price deflator
GEF	Global Environment Faculty
GHG	Greenhouse gas
GP	general practitioner
GVW	gross vehicle weight
HBJ	Hazira-Bijaipur-Jagadishpur (Indian gas pipeline)
HC	Hydrocarbons
HMGN	His Majesty's Government of Nepal
ICR	Implementation Completion Report (World Bank)
IFC	International Finance Corporation
I/M	inspection and maintenance
IMF	International Monetary Fund
IPP	independent power producer

IRR	internal rate of return
JCC	Japan crude cocktail
JICA	Japan International Cooperation Agency
JPY	Japanese yen
kPa	kilopascals (a measure of pressure)
LECO	Lanka Electricity Company
LPG	liquid petroleum gas
LRMC	long-run marginal cost
MC	motorcycle
MEC	maximum effective capacity (of a rail line)
MEIP	Metropolitan Environmental Improvement Programme
mmBTU	Million British Thermal Units
MON	motor octane number
MoU	Memorandum of Understanding
NBRO	National Building Research Organisation
NPV	net present value
NRs	Nepali rupees
NSL	National Security Levy
OCCT	open cycle combustion turbine
OECD	Organisation for Economic Cooperation and Development
OLS	ordinary least squares
OTEC	ocean thermal energy conversion
PCU	passenger car unit
PG&E	Pacific Gas & Electric (California)
PIA	project impact area
PICD	port inland clearance depot
PLF	plant load factor
PPP	purchase power parity
psi	pounds per square inch
RDA	Road Development Authority
RFP	request for proposals
RON	research octane number
RVP	Reid vapour pressure
SCADA	Sone Command Area Development Authority (Bihar, India)
SCF	standard correction factor
SLR	Sri Lanka Railways
SLCTB	Sri Lanka Central Transport Board
SURE	seemingly unrelated regression estimate
T&D	transmission and distribution
TEL	tetra ethyl lead
TSPC	Transport Studies Planning Centre

TSSS	Transport Sector Strategy Study [World Bank]
UNFCC	United Nations Framework Convention on Climate Change
URT	upper respiratory tract
USAID	United States Agency for International Development
VLCC	very large crude carrier
VOC	vehicle operating cost
WASP	Wien Automated System Planning (model)
WP	Western Province
WPI	wholesale price index

Foreword

This book deals with energy, which is considered the lifeblood of modern societies. The energy sector has been the dynamic engine of development and growth in recent times. In addition to this dominant economic contribution, the production and use of energy also has important environmental and social implications. Therefore, sustainable energy development requires balanced and careful consideration of key economic, social and environmental issues by all stakeholders -- including decision makers, producers, consumers, policy analysts and researchers.

This volume is the second in the MIND series which examines the sustainability of growth. The first book (*The Sustainability of Long Term Growth* by Munasinghe et al., 2001, Edward Elgar, Cheltenham, UK) has been well received. The third volume (*Macroeconomic Policies for Sustainable Growth* by Munasinghe et al., 2005, Edward Elgar, Cheltenham, UK) presents country-specific case studies of Brazil and Chile.

The present book looks at the subject of energy in its many dimensions, and serves as an excellent input for the resolution of many energy and environmental issues. The authors have produced a comprehensive and lucid book, which demonstrates a thorough grasp of the complex relationships involved. A multidisciplinary approach is adopted, which is well suited to analyzing sustainable development issues. Emphasis is given to the valuation of environmental and social impacts, such as cost of health damage due to air pollution, and economic and environmental cost of urban congestion and fuel price distortions. Although the book contains highly technical material, it has been presented in a very readable manner, even to the average reader.

The issues and problems of sustainable energy development are common to most countries. Therefore the lessons learned from the case examples in Sri Lanka and other countries, offer good insights to energy planners not only in Sri Lanka, but also elsewhere. Some of the specific problems and issues covered include, *inter alia* :

a) Choice of power generation technologies such as hydro, thermal, LNG, coal, renewable sources etc., and the implications of different choices.
b) Economic, environmental, social, political and technical dimensions of power plant siting.
c) Challenges of institutional reform involving the transformation of monolithic, state-owned energy organizations such as 'Ceylon Electricity Board' and 'Ceylon Petroleum Corporation', into decentralized entities with private sector participation.

d) Declining investments in the public transport system, coupled with rapid increases in private vehicle fleets and higher consumption of imported fuels.
e) Implications of global climate change for national strategies of sustainable energy development.

The authors demonstrate their command of not only conventional techno-economic issues covering electric power and fossil fuels, but also recent developments in areas like renewable energy, climate change, and the environment. Dr. Peter Meier is an eminent energy expert, and frequent visitor to Sri Lanka. He has had extensive field experience on energy and environmental issues in many developing countries. He has collaborated closely with Professor Munasinghe for over two decades and published widely on energy. He has served as a senior consultant to the World Bank, as well as to many other donor agencies, governments, and private sector energy companies. Dr. Meier has also held senior teaching appointments in several universities, and presently heads his own consultancy company.

Professor Mohan Munasinghe is a world-renowned authority on both energy and sustainable development. His comprehensive knowledge of the energy situation in Sri Lanka draws on his current experience as my Honorary Chief Energy Advisor, as well as his former role as Senior Energy Advisor to His Excellency President Jayewardene of Sri Lanka (1982-87). This book has also benefited from his exposure to a wide range of development activities, while serving as a senior manager at the World Bank, where he successively headed major departments covering not only energy and infrastructure, but also environmental policy. His mastery of the subject matter in this volume is confirmed by his authorship of a large number of other books and papers on energy and environment, as well as the receipt of the highest professional awards in the field, including two major prizes from the International Association of Energy Economics (IAEE) and the United States Association of Energy Economics (USAEE).

One major thread running through the book is a sense of detached objectivity, coupled with the authors' genuine desire to share the depth of their knowledge and experience without reservations, for the benefit of the world community. I recommend this book to political leaders, policy analysts, energy planners, researchers, academics and concerned individuals associated with energy and development, both in Sri Lanka and elsewhere.

Karu Jayasuriya MP
Minister of Power and Energy
Colombo, Sri Lanka
March 2004

1. Introduction

Sri Lanka's energy sector stands at a crossroads. Historically for power generation the country has been almost entirely dependent upon hydro-electricity. Since independence in 1972, it has relied largely on a Government-owned, vertically-integrated monopoly – the Ceylon Electricity Board (CEB) – for power generation, transmission and distribution. Until the early 1990s, this served the country well but, over the following decade, the power sector has been embroiled in a series of acrimonious disputes – over the choice of thermal technology and fuels; over power plant siting; and over reforms of the institutional arrangements to bring greater efficiency to the sector and to deal with the financial implications of a greater involvement by the private sector. The result has been paralysis in decision-making, financial distress, and significant power cuts (notably in 1996 and again in 2001/02) leading to an increasingly severe impact on economic growth.

The transport sector faces its own range of issues, no less intractable. Traffic congestion in Colombo has risen sharply over the past few years, and congestion in the Port of Colombo needs to be reduced if Sri Lanka is not to lose its role as a regional trans-shipment port for containers. While it may be true that congestion has been aggravated by security measures in Colombo which have restricted traffic in many key roads, this cannot justify delays in implementing a range of environmental measures – notably the introduction of unleaded gasoline – that were constrained by a combination of spurious objections and bad pricing policy: no country in Asia has successfully introduced unleaded gasoline following Sri Lanka's earlier approach of pricing unleaded gasoline higher than leaded gasoline. However, following pressure from the Civil Society, leaded gasoline has been phased out completely.

Indeed, the need for comprehensive reform of fuel taxation is urgent. The distortions in fuel taxation policy have led to widespread conversion of gasoline vehicles to LPG – notwithstanding that LPG is by far the most expensive transportation fuel when valued (as economic analysis requires) at the border price. The discrepancy between taxes on diesel and gasoline fuels has led to the use of large numbers of diesel vehicles.

Box 1.1 Sri Lanka's energy sector institutions

Ceylon Electricity Board (CEB): Government-owned, vertically integrated utility, owns and operates all major hydroelectric stations, some thermal plants, the entire transmission grid, and most distribution facilities. CEB serves 89% of all customers, and in 2002 sold 4,691 GWh.

Lanka Electricity Company (LECO): Government-owned distribution company, established in 1983, which took over municipally-owned distribution companies in the Colombo suburbs at a time when they were experiencing T&D loss rates in excess of 50%. Mandated to operate on a commercial basis, by the late 1990s loss rates had declined to around 10%: in 2002 LECO reported losses of 6% (of energy purchased) from CEB. LECO owns the distribution network in its service territory, and serves about 11% of all customers. In 2002, LECO sold 763 GWh.

Ceylon Petroleum Corporation (CPC): Government-owned company established in 1961 (following nationalisation of three oil majors), held the oil import, refining and distribution monopoly, but now undergoing reform and restructuring. Its Sapugaskanda refinery was commissioned in 1969. The nylon plant was divested in 1992; the lube business in 1994; the LPG business in 1995; and the bunkering business in 2002. In 2004, a new company, *Ceypetco*, was formed to CPC's fuel marketing, refining and import operations.

Lanka Indian Oil Company (LIOC): subsidiary of the Indian Oil Company, owns and operates some 100 filling stations, and leases the Trincomalee tank farm from CPC. LIOC started operations in Sri Lanka following agreement with the Government of Sri Lanka in June 2002; under terms of the agreement, permitted to import products directly. Another entity is being selected by the Government of Sri Lanka to become the 'third player' in the oil sector.

Ministry of Power and Energy (MOPE): oversees the Government-owned entities (CEB, LECO and CPC).

Energy Conservation Fund (ECF): operates under MOPE, with a mandate to promote energy efficiency. Publishes an annual energy balance.

Public Utilities Commission (PUC): Established under the October 2002 PUC Act, will initially regulate the electricity industry, but in the future will also regulate other entities such as water supply and petroleum.

Shell Gas Lanka, Gas Auto Lanka, Lanka Gas, and Mundo Gas: Companies that import and retail LPG. When the Colombo Gas Company was privatised in 1995, the Government allowed a five-year period during which Shell Gas Lanka had an effective monopoly, until market liberalisation brought new entrants after 2001.

Energy Supply Committee (ESC): established in March 2002 to oversee and coordinate procurements required to overcome the electricity crisis of 2001/2002. Also has a mandate for energy planning, and to issue licences under the Electricity Act.

This has led in turn to increased diesel fuel imports, even while naphtha (the feedstock for gasoline) has been in surplus and exported at a loss. As we show in this book, the result is a system of cross-subsidies the beneficiaries of which are rarely the poor. Moreover there is no reason why tax reforms cannot be undertaken in a revenue-neutral way.

Renewable energy is subject to its own controversies. While programmes for village hydro and solar homes have been successful, energy generation at the CEB wind demonstration project has been disappointing with much lower than expected wind speeds. The grid-connected small hydro programme has attracted a number of private developers and good private sector interest, but the pricing scheme remains problematic as CEB and the developers struggle over definitions of avoided costs.

Resolving these issues in the face of uncertainty about economic growth and the world oil price is not easy, and predictions about the consequences of geopolitical events are extremely difficult. Following the terrorist attacks on Colombo's international airport in July 2001, the Sri Lanka tourist industry suffered a catastrophic contraction, which had severe impacts on both the balance of payments and the domestic economy and employment in 2002. In 2003 tourist arrivals and the economy have picked up following resumption of the Norwegian brokered peace process, yet there remain many uncertainties. The possibility of the oil price increasing to unprecedented levels in the aftermath of further turmoil (or revolution) in Saudi Arabia is ever present.

Over the long term there are further issues of uncertainty about the impact of climate change, and the extent to which Sri Lanka could benefit from its low per capita emissions under a global carbon-trading regime. But in Sri Lanka even this debate has been characterized by bizarre discussions about the potential utilization of liquid natural gas (LNG) for power generation. Since the price of LNG is directly linked to that of oil, the risks of a commitment to LNG would be very high; but while local air and greenhouse gas (GHG) emissions would certainly be less than those of coal-fired generation, they are still orders of magnitude larger than those from dendro-thermal plants – whose fuel costs are not subject to the uncertainties of the turmoil in the Middle East and of global recessions.

1.1 OBJECTIVES

Over the past few years we conducted five detailed studies of Sri Lanka's energy sector that cover many of these challenges: a study of GHG emission mitigation options for the power sector (Fernando and Munasinghe, 1998; Meier and Munasinghe, 2003c); a study of technology choices for the power sector (Meier and Munasinghe, 2003a); a study of the environmental implications of transportation sector options (Meier and Munasinghe, 2003b); and a study of renewable energy (Meier, 2003c). Our purpose here is to integrate these studies into a single volume – whose main objective is to contribute to the resolution of the many current energy and environmental policy issues by an objective analytical approach using the tools of environmental economics and cost-benefit analysis.

How to deal with risk and uncertainty, how to quantify and display trade-offs, and how to identify robust policy options are themes that run throughout. Fortunately it turns out that environmentally sustainable solutions to Sri Lanka's energy problems are often the most robust with respect to exogenous uncertainties.

This quantitative focus also serves as one of the main boundaries of the study – for there exists a range of issues, particularly regarding power sector reform, that lie clearly outside such a boundary. Thus we do not examine the regulatory, institutional, managerial and labour relations dimensions of power sector reform, though we do look at the project financing implications of greater private sector involvement, and at the technical consequences of potential reforms.

We also limit our focus to commercial energy, and in particular we exclude the domestic fuelwood sector. But we do evaluate the role of off-grid renewable energy, such as solar home systems (SHS) and village micro-hydro, for providing electricity in rural areas.

1.2 THE MAIN ISSUES

The Choice of Power Generation Technology

With its hydro potential largely developed, Sri Lanka's transition to thermal generation is inevitable. While renewable energy and demand-side management have significant roles, neither is likely to be able to eliminate the need for fossil-based generation. Given significant increases in the load, the potential emerges for exploiting the scale economies inherent in larger thermal generation units (Box 1.2). Yet these scale economies also give rise to a lock-in effect: if, for example, coal is chosen for the first new 300 MW baseload unit, it requires a commitment not just to that first unit, but to at least an additional two units. LNG is subject to even greater lock-in effects.

Box 1.2 Scale economies in thermal power generation

The figure below shows the decreasing capital cost per kW as a function of increasing net plant output, for coal and combined cycle gas turbines (CCCT). These costs exclude fuel handling infrastructure costs – single point mooring for oil, or coal jetty for coal; were these included, the indicated scale economies would be even greater. They are as estimated by various engineering consultants for projects in Sri Lanka.

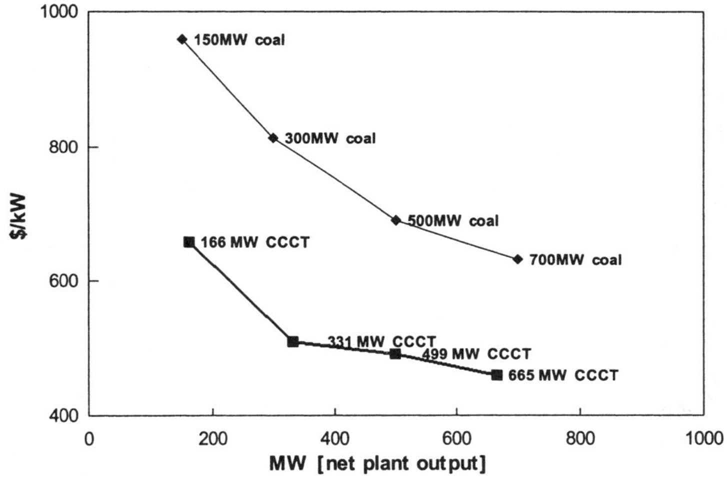

Source: Electrowatt Engineering (1996)

Even the environmental trade-offs may be difficult to resolve: while oil avoids the particulate emissions and solid-waste generation of coal, the cheapest heavy fuel oil has three to four times the uncontrolled sulphur emissions of a coal plant. At the same time, LNG – certainly the most attractive fossil fuel option from an environmental point of view – is presently also by far the most expensive. However, the global LNG pricing regime is undergoing significant changes, and there are indications that LNG prices may become decoupled from crude oil prices in the future.

Oil-fired diesel and combustion turbine units can be built in smaller increments, avoiding such lock-in effects. Yet they expose Sri Lanka to other risks, notably the volatility of price and susceptibility to geopolitical supply disruption: in the current climate of global instability, any number of scenarios could lead to sudden increases in the price of oil

– particularly a violent challenge to the current regime in Saudi Arabia or further escalation of the Israeli–Palestinian conflict. Even if central Asia has the reserves to make up part of the global demand shortfall, it will take many years before this can be brought to world markets. Moreover, China's increasing dependence on oil imports will add further long-term pressures on oil prices. In short, commitment to oil-fired power generation carries high risks, which should be properly hedged by diversification of supply.

Power Plant Siting

NIMBY ('not in my backyard') is alive and well as the paradigm for power plant siting in Sri Lanka. The seemingly unending controversy over the site of the coal-fired power plant illustrates the difficulties of managing public debate into constructive channels. The original choice of site at Trincomalee was soundly based in economics – notably the desire to exploit the excellent natural harbour that would permit transportation cost savings of up to US$3/tonne (since coal could be imported in the very largest of vessels without trans-shipment). Although it is tempting to conclude that the failure of the plant to receive the necessary permits was a victory for the environmental lobby, in fact the main environmental impacts cited could have been mitigated at little additional cost. For example, the cooling water could just as well have been discharged into the deeper waters of Trincomalee Bay, where upwelling effects would have been beneficial to fisheries, as it was by discharge into the shallow bay actually proposed, where it was demonstrably deleterious (Meier and Munasinghe, 1994). In fact, given the security situation on the east coast (areas controlled by the Tamil Tigers), even had the environmental permits been given, financing as a public sector project would have been very difficult – and financing as a private power project even more so.

Subsequently a site on the south coast was proposed (and then rejected), followed in the late 1990s by another site at Puttalam, north of Colombo, which was strongly opposed by environmentalists, so no firm decision was made by the Government. With the resumption of peace talks with the Tamil Tigers in 2002, Trincomalee was again proposed as a long-term solution, to be implemented as an IPP: however, recognising that this cannot not occur absent a permanent peace settlement, oil-based generation is under renewed consideration for the short term.

These are not the unique problems of Sri Lanka, or even of developing countries in general, for poor power siting policy has plagued most developed countries as well. There is no better example than the Shoreham Nuclear Power Plant on Long Island, New York – where only after the expenditure of US$ 5 billion for its construction, com[illegible] 1985, was it decided that the site was in too populated an area to permit safe emergency evacuation, and the plant was abandoned in 1986. But as at Puttalam, this argument was merely the convenient crutch of politicians who saw votes in opposition to nuclear power: the population distribution around the site was perfectly known well long before construction began.

The result is that years after a coal plant was originally proposed, paralysis and indecision reigns as the same issues are debated again and again, with no resolution in sight. Meanwhile, short-term *ad hoc* measures are taken to fill the growing power gap – whose environmental impacts are demonstrably far worse than those of a coal plant. As we shall discuss below, it is hard to conceive of a worse site for the barge-mounted power plant (see below for details, Table 2.5) than Colombo harbour: the plant's emissions of sulphur and particulates – regularly blown into the most densely populated urban area of the country by sea breezes – have a demonstrably greater impact on Colombo's population than any coal plant. Yet the environmental impact of the barge-mounted plant attracted neither the interest nor protest of non-governmental organisations (NGOs).

Thus the present outlook for additional generating capacity to meet growing demands is poor. The 150 MW Upper Kotmale hydro plant, which in the 2003 CEB Plan is to be available in 2009 (a plant first included in the 1994 CEB plan for commissioning in 1999!), remains dormant, despite a 2002 cabinet decision to proceed with implementation: as of early 2004, contractors had yet to be mobilised, and completion by early 2009 is not possible. The coal project, assumed to be available by 2008, and now planned to be implemented as an IPP, is stuck in a seemingly endless pre-qualification process, and its target completion date is similarly unachievable. The 300 MW Kerawalapitiya project, first included in the 1998 CEB plan for completion in 2003, was also mired in a lengthy selection process, taking one year to evaluate pre-qualification applications. Even if financial closure could be reached by end 2004 (a heroic assumption), the 2003 CEB Plan commissioning date of January 2006 cannot be reached. The prospects are therefore for continued power shortages, even in average hydro years, which almost inevitably will be met by additional stop-gap projects running on auto-diesel, about the most expensive possible way of generating power.

Managing the Transition

A third major issue concerns the institutional arrangements. The power sector is currently undergoing major reforms. The main objectives of these reforms are to lower the price to consumers while ensuring a high level of services and supply reliability, and to sustain an adequate level of investment in the power sector. The current vertically integrated monopoly structure will be unbundled and separated to form one generation, one transmission and several distribution entities, with the transactions based on a single buyer model. An independent regulatory commission will be set up to regulate the power sector.

Two acts were enacted by parliament in October 2002: the Electricity Reform Act and Public Utilities Commission Act. Based on these two acts, CEB and LECO are to be reorganized to form several new entities by mid-2004. The Public Utilities Commission of Sri Lanka was also established by appointing the Commissioners, and is expected to function fully before the establishment of the new sector entities.

A major objective for reform and restructuring of the power sector is the mobilization of private capital. But the transition from public to private financing has profound implications for the cost of power. At the heart of this debate is concessionary financing. One view is that concessionary financing should be reserved for social sectors for which private finance is not available – which is the position of the Government. However, this has two main consequences: a higher cost of power (because of higher financing costs); and the continuing need for sovereign guarantees, for it is hard to see that private power projects would not require these for the foreseeable future (as has been the case in all major IPPs in Sri Lanka to date). In any event, concessionary financing is available only to public agencies, and there are doubts whether CEB is capable of undertaking a major coal-fired project. But that is an argument for speeding up reforms in CEB, not an argument against concessionary finance.

Transportation Congestion and the Environment

The challenges for the transportation sector are no less significant. While traffic conditions and urban air quality in Colombo have yet to deteriorate to levels encountered in many other Asian cities,[1] particularly over the past five years there has been a significant increase in road traffic congestion in Colombo, as sustained economic growth and lagging investment in public transportation have led to increased vehicle ownership and corresponding growth in transport fuel consumption.[2]

Table 1.1 shows survey results taken on the seven main radial roads from the centre of Colombo for the morning peak in 1997, 1999 and 2001. Over the four years 1997–2000, speeds on all roads have fallen by at least half; moreover, while in 1997 peak conditions were observed only for roughly two hours during the morning peak, by 2001 these conditions extended until mid-afternoon (in the inward direction); and similarly in the outward direction, the congested period now starts in the mid-afternoon. As discussed below, such dramatic reductions in network speeds result in significant increases in local air emissions, because fuel efficiency decreases non-linearly with speed – especially under stop–go driving conditions on highly congested roads.

Table 1.1 Journey time survey results (average speeds, km/hour): inward direction, morning peak

	1997	1999	2001
Colombo–Galle	28	18	8
Colombo–Ratnapura	32	16	6
Colombo–Kandy	23	22	11
Colombo–Puttalam	18	16	7
Jayawardanapura–Kollupitiya	23	17	9
Colombo–Horana	27	17	8
Wellampitiya–Kaduwela	28	18	9

Source: Transport Studies Planning Centre, Ministry of Transport

Sri Lanka expects continued strong growth in transportation demand that inevitably accompanies economic growth. As shown in Figure 1.1, liberalization of the economy in 1977–78 was followed by an extraordinary increase in the vehicle fleet as physical import controls were relaxed and decades of pent-up demand was realised.

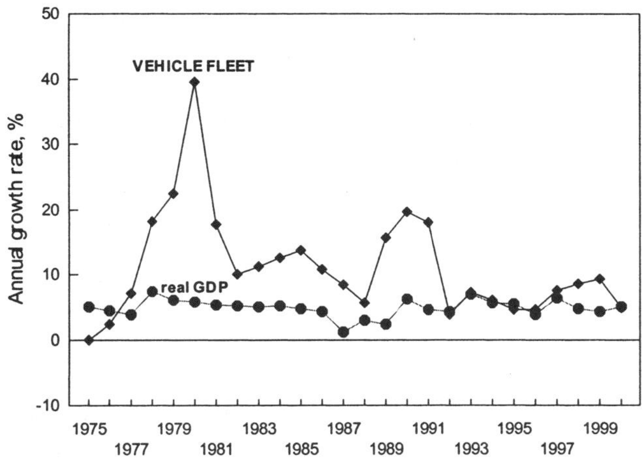

Figure 1.1 Growth of the vehicle fleet and GDP

A second vehicle fleet growth boom – albeit not quite as sharp as the previous one – occurred in 1989–91, after the end of the civil disturbances of 1986–89. Since the early 1990s, growth of the fleet has tracked GDP growth very closely.

Such robust economic growth inevitably leads to an increased burden of local air emissions. Figure 1.2 shows the projection of transport-sector emissions made in the mid-1990s by the Clean Air 2000 Action Plan (MEIP, 1993). This Plan expected air emission increases of 57–68 per cent over the decade 2000–10.

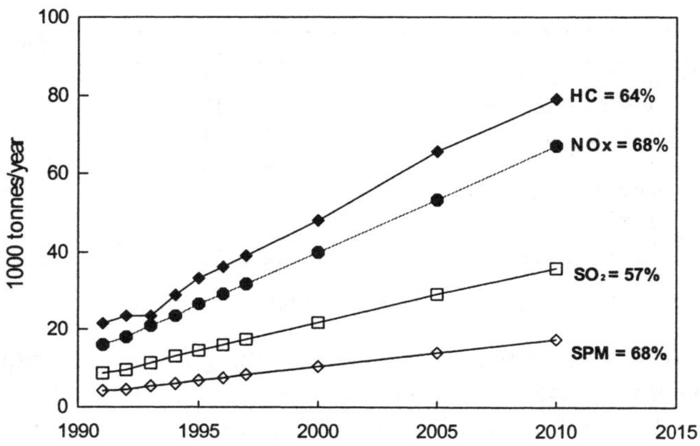

Figure 1.2 Projected transport sector local air emissions

Without appropriate and timely policy interventions, the present trends will inevitably lead to a much higher level of degradation both in the urban environment and the quality of life as journey times increase. Previous estimates of the total costs of congestion (and accidents) in Sri Lanka – but excluding the local environmental externalities – amount to roughly 0.1 per cent of GDP,[3] comparable to that of other Asian countries.[4]

It therefore seems prudent to take action before the situation becomes significantly worse. First, as is true for many environmental issues that involve externalities not captured by markets and prices, prevention of further deterioration through proper planning is often cheaper than ex-post clean-up. Moreover, vested interests in some unfortunate present practices will become more entrenched the longer they are permitted to continue – for which a good example is the practice of on-street parking on many of Colombo's arterial roads, a major cause of congestion externalities.[5]

Second, many potential policy measures have long lead times. For example, since an outright ban on two-stroke motorcycles is not practicable, efforts to lower the proportion of these vehicles in the fleet mix by banning further imports (or by tax disincentives) would necessarily require many years before all of the existing fleet of two-stroke motorcycles is retired.

The Implications of Climate Change

Munasinghe (2001) uses the 'sustainomics' trans-disciplinary framework to show how climate change and sustainable development interact in a circular fashion, especially in the longer term. Alternative development paths will certainly determine greenhouse gas (GHG) emission levels that would affect future climate change, and have implications for mitigation strategies as well. In turn, climate-change vulnerability, impacts and adaptation will influence prospects for sustainable development in the future. In this book, however, we focus only on the narrower questions of short to medium term GHG emission reduction strategies in the energy sector.

Sri Lanka's present emissions of GHGs are modest by international standards, but are expected to quadruple over the next decades as the power sector shifts increasingly to thermal generation, and emissions from road traffic see an inevitable rise.

Nevertheless, despite these increases, by 2020 Sri Lanka's per capita (or per GDP) emissions will still be a small fraction of the world average. Sri Lanka therefore stands to benefit significantly from any global carbon trading regime or from projects under the Clean Development Mechanism (CDM) of the Kyoto Protocol. However, the economic benefits to Sri Lanka from carbon trading will depend critically on sound planning, and on the highly uncertain value of carbon offsets.

It has been argued that the higher costs of LNG could be financed by such carbon trading benefits. But that brings little net benefit to Sri Lanka if these funds are simply used to pay for high-cost projects. There is a much greater potential for net benefits from renewable energy projects where Sri Lanka has a real comparative advantage, such as dendro-thermal power and small hydro.

1.3 PAST WORK

Although many detailed project level environmental impact statements have been prepared since the late 1980s (starting with the environmental studies for the proposed coal-fired Trincomalee plant), Meier and Munasinghe (1994) made the first attempt to systematically analyse the environmental dimensions of Sri Lanka's power sector development. Subsequently they examined the GHG emission consequences of power development strategies (Meier and Munasinghe, 1995). A comprehensive study of GHG emissions by the Ministry of Forestry and Environment appeared in 1998 (Ratnasiri, 1998). Proposals for major projects in the power sector have benefited from comprehensive environmental assessments since the mid-1980s, and for many years the annual generation expansion plans of the Ceylon Electricity Board have included presentations of the air emission consequences (for example, CEB, 1998b, 1999). Several studies have also appeared from NGOs, notably LIFE (1998), that examined the environmental consequences of the 1996 power shortages.

On the other hand, the environmental aspects of Sri Lanka's transportation sector have received much less attention. The work in the early 1990s that led to the adoption of the Colombo Clean Air 2000 programme (MEIP, 1994) focused mainly on emissions, but lacked systematic cost–benefit analysis. Subsequent analyses of measures to reformulate fuel (lower lead in gasoline, lower sulphur in diesel) were limited to financial analysis of the impact on the refinery. After intense criticism of the first Environmental Impact Assessment of the proposed Colombo–Katunayake expressway, a new assessment in 1997 was the first attempt by the Road Development Authority (RDA, 1997) to monetize local environmental impacts and incorporate them into cost-benefit analysis. However, as we will see in Chapter 6, this assessment was not without problems. Transportation sector emissions were included in the GHG emissions inventory (Ratnasiri, 1998), and selected transportation measures were examined in Meier et al. (1997), but to date there has been no systematic examination of the GHG emissions consequences of a broader range of policy options in the transport sector.

An important study of congestion costs was completed in 1999, but this did not extend to valuation of the relevant environmental externalities (Jayaweera, 1999). A comparison of the environmental consequences of selected policy options in the road transport sector was published in 2000, but the focus was limited to the trade-off between costs of policy implementation and emissions, and did not extend to benefit valuation (Chandrasiri, 1999). The first rigorous study of health effects of vehicle pollution in Sri Lanka also appeared in late 2000, but it too was limited in scope, and quantified health damages only for a small group of highly exposed occupations (Chandrasiri and Jayasinghe, 1998). Another study of hospital attendance for asthma suggested a strong correlation with air pollution levels in Colombo (Senanayake et al., 1999).

1.4 SCOPE AND OUTLINE

We begin, in Chapter 2, with a review of recent trends and the present situation. We review in particular the air quality trends in Colombo, the evolution of the transport sector, and the main features of the power sector. The emphasis on Colombo follows from the urbanization pattern of the country: in 1998, the Greater Colombo Metropolitan Region (GCMR) area had some 4.8 million inhabitants,[6] about 25 per cent of the total 19.4 million population of Sri Lanka.[7] Kurunegala district has the second highest population outside the GCMR, with 1.5 million inhabitants.[8] It is estimated that 1.7 million persons cross the Colombo city boundaries every day, of which 90 per cent travel by road (Jayaweera, 1999).

Chapter 3 examines the valuation of air emissions. As noted, two recent studies in Sri Lanka have monetized air emission damages, both based on the benefit transfer method: the 1997 environmental impact statement for the Colombo–Katunayake expressway and a study of the impacts of the 1996 power shortages. These two studies are compared to estimates in other Asian cities, and to the values derived in recent World Bank studies.

Sri Lanka has not been blessed by nature with fossil fuel resources, which must therefore be imported. Unfortunately, assumptions about fuel cost in past analyses often lack consistency. This is not merely a matter of the uncertainty of predicting the future world oil price, but in assumptions about relative prices that often have no (or little) foundation – for example, that coal prices are largely independent of oil prices. Thus Chapter 4 presents a framework for assessing the relative prices of imported fuels (based on a statistical analysis of the past 30 years). We make no claim that we can make better forecasts of future energy prices than others. But we do claim that the economics of refining, and the substitutability of heavy fuel oil and coal, impose clear constraints on relative prices.

How to make technology and fuel choices in the power sector then follows in Chapter 5. Here we summarise the main findings of the detailed report on the Sri Lanka Electric Power Technology Assessment

(SLEPTA), (IDEA, 2000; Meier and Munasinghe, 2003a). We note in particular the differences between economic and financial analysis, and how results depend upon financing packages.

Chapter 6 develops the methodology for including environmental costs and benefits in transportation sector cost–benefit analysis: we illustrate this in detail for the case of congestion pricing; we also summarise the results for the other transportation sector options examined in the detailed transportation sector study (Meier and Munasinghe, 2003a), and review the estimates of fuel-price, vehicle, and km-cost elasticities.

The consequences of fuel price distortions are examined in Chapter 7. We begin with a discussion of the conversion of gasoline passenger cars to LPG – a textbook example of a development driven entirely by misguided tax policy to the clear detriment of economic efficiency. We then examine taxation of diesel and gasoline fuels, and show that both equity and economic efficiency would be improved by an equalised tax which avoids the inequitable cross-subsidies implicit in the existing system – namely from lower-income motorcycle owners to the road freight industry and upper-income groups who have moved increasingly to diesel sports utility vehicles (SUVs).

Chapter 8 discusses the question of phasing out leaded gasoline. We show that many of the arguments heard in Sri Lanka over the past decade against this policy have no merit. We draw on the experience of other countries (particularly Eastern Europe) where the objections would have much more merit than in Sri Lanka, but where unleaded gasoline was successfully introduced in the 1990s.

Sri Lanka's options for reducing GHG emissions are presented in Chapter 9. We derive combined avoided cost curves (the familiar 'staircase' presentations of the reductions in GHGs at various costs of avoided carbon), based on a detailed analysis of power sector options in which the baseline scenario is perturbed by individual mitigation measures (such as dendro-thermal, LNG or wind plants). For the transport sector we examine the GHG emission implications of four possible scenarios of long-term development – ranging from the extremes of 'American dream' (with auto ownership stabilising at 800 vehicles per 1000 households) to 'Singapore controls' (with auto ownership stabilising at around 300 vehicles per 1000 households).

In Chapter 10 we develop a real options framework for dealing with the uncertainties of a potential global carbon-trading regime (Fernando and Munasinghe, 1998). Unlike a number of recent papers that have presented generalized and highly aggregated discussions of options theory, our framework draws more narrowly on so-called real options developed in the financial markets, and is applied to specific project choices as may be available to the Government.

Renewable energy for rural electrification is examined in Chapter 11. Access to electricity and poverty reduction are closely linked, and by 2003 only about two-thirds of Sri Lanka's households were electrified. We examine first the problem of estimating the benefits of electrification using changes in consumer surplus, followed by a closer look at two off-grid options for rural electrification: solar homes and village hydro.

Finally, in Chapter 12 we draw together our recommendations. The emphasis here is to draw the appropriate lessons from Sri Lanka's disappointing experience over the past decade. To be sure, in a democratic political process there will always be dissent, and the hesitancy of decision-makers to make firm decisions is perhaps understandable in the face of a debate that is largely qualitative and *ad hoc* (as shown by the complete lack of interest on the part of many environmentalists – though with some notable exceptions – to engage in any real discussion of trade-offs). However, there is an alternative to indecision and policy-making paralysis, for energy and environmental trade-offs can be quantified (and in some cases monetized), and this surely provides decision-makers with greater insights about the consequences of alternative options. Nevertheless, while sector reform and restructuring is a necessary step, one should be under no misapprehensions about the most fundamental constraint to attracting capital, and particularly foreign investment, into the energy sector for sorely needed infrastructure projects, namely the permanent resolution of the ethnic conflict.

NOTES

1 The World Bank Transport Sector Strategy Study (World Bank, 1996b) notes that "...Sri Lanka's air pollution problem appears to be currently at manageable levels. Yet based on projected rates of economic, demographic and vehicular growth, the air pollution situation in Colombo could come to resemble that of more severely affected Asian countries. The Clean Air 2000 action plan notes that the transport sector is the biggest contributor of air pollutants in the Colombo metropolitan area...thus it is clear that the transport sector's contribution to environmental problems overall in the Colombo metropolitan area is high."

2 Comparisons with petroleum product forecasts made a decade ago to actual figures are instructive. For example, the 1992 forecast by the Ceylon Petroleum Corporation (CPC) for the year 2000 was for total diesel consumption of about 900,000 tonnes, as against an actual diesel consumption of some 1.3 million tonnes. However, most of the forecast error was for power generation: in 1992 the forecast for the year 2000 diesel consumption for power generation was estimated at 135,000 tonnes as opposed to an

actual figure of some 480,000 tonnes. In the early 1990s it had been assumed that a proposed 300 MW coal-fired plant would be in operation by the mid 1990s, which has not in fact occurred. The result has been a sharp increase in self-generation as hotels, industry and commercial enterprises have responded to endemic power cuts by increasing self-generation. For the transportation sector, the year 2000 forecast for auto-diesel was 757,000 tonnes, as against our estimate of the actual consumption of around 850,000 tonnes.

3 World Bank (1996b), para. 5.5. The estimates cited in that study range from Rs 550 million to Rs 778 million.

4 For example, in Thailand, congestion costs in Bangkok were estimated at US$ 272 million/year, representing 0.2 per cent of Thai 1994 GDP of US$ 1.28 billion (World Bank, 1994).

5 Kumarage (1996) gives an excellent example of externalities caused by inappropriate land use and related parking problems in Dehiwalla on the Galle Road (one of Colombo's seven major radial arterial roads, whose congestion problems are discussed in Chapter 6). After grape sellers established themselves there, average speeds dropped by about 4 km/hour, as drivers stopped at the kerbside to make purchases. This speed reduction on a road that carries 50,000 vehicles per day results in increased gasoline consumption of about 500 litres/day, imposing costs to the economy of some Rs 2 million/year, equal to about Rs 20,000 for each of the approximately 100 grape sellers.

6 Comprising 2.2 million in Colombo district, 1.6 million in Gampaha, and 1 million in Kalutara. These three districts comprise the Western Province, which most studies equate to the GCMR.

7 Preliminary estimate for the year 2000 as reported by Central Bank of Sri Lanka (2000).

8 Comparisons with Bangkok are instructive: in 1991, the Bangkok metropolitan area accounted for one-sixth of Thailand's total population, and had 15 times the population of the second largest city, Chiang Mai (World Bank, 1994).

2. The present situation

This chapter reviews the main historical trends and assesses the present situation. We begin with a review of local air quality in Colombo. This is followed by a review of the transport sector, which saw fundamental changes following liberalization of the economy and the lifting of import controls in 1978. Finally, we summarize the main features of the power and petroleum sectors.

2.1 AIR QUALITY[1]

Compared to the air quality data available in other Asian cities, that for Colombo is quite limited. Systematic monitoring started in 1989 under the auspices of the Environmental Division of the National Building Research Organisation (NBRO).[2] A first study examined 49 locations in the city of Colombo for sulphation rate and dustfall, followed by a second programme at seven representative locations in 1992–93.[3] In late 1996, NBRO started a long-term monitoring programme at two Colombo locations – at Fort, in the centre of the city, and a site in Colombo 7 (Bauddaloka Mawatha, a mainly residential area) as a background station. These sites have provided PM-10 measurements once in four days, with continuous monitoring of NOx, CO, SO_2, and ozone (plus corresponding meteorological data). This is the most important source of Colombo air quality data.

Has air quality in Colombo in fact deteriorated, and if so to what extent can such deterioration be linked to vehicular emissions? Unfortunately, because of differences in sampling methods and locations, the new data from 1996–2000 is difficult to compare with the earlier data from 1989 and 1992–93. However, the new (and consistent) data set since late 1996 does permit some preliminary conclusions.

Figure 2.1 shows monthly SO_2 averages from 1996 to 2000. We note the seasonality of the data: the highest concentrations are in November–January, which reflects the impact of fumigation conditions during the NE monsoon – when the prevailing wind direction is exactly opposite to the sea breezes that have W to SW direction, typical during the late afternoon–evening hours.

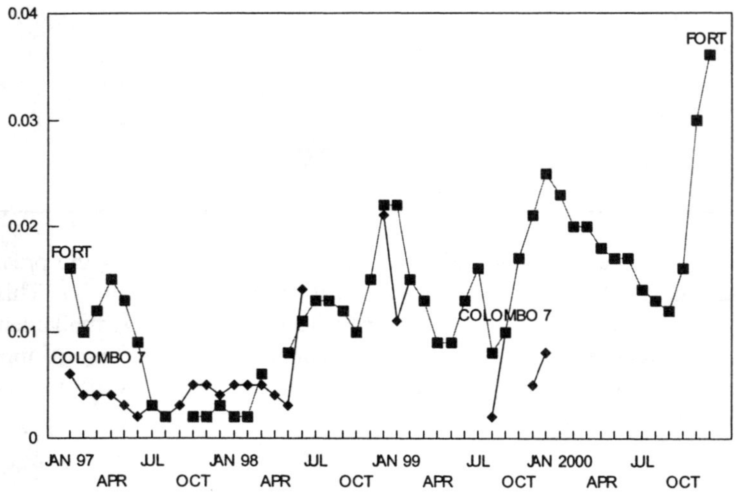

Figure 2.1 Monthly averages for SO_2

Although the NBRO Fort station data is unbroken over the four-year period, there are extensive gaps in the available data for the Colombo 7 location, so it is difficult to make judgements about the extent to which the Fort location – clearly a 'hot spot' – can be used to gauge average population exposures

Nevertheless, the Fort data do reflect the growth in vehicular diesel emissions, the main source of SO_2: the winter peaks get worse each year (except 1997 – whose winter values show no peak at all),[4] most notably in 2000 – which is almost certainly due to the start-up of the 60 MW barge mounted power generation project in Colombo harbour.

Another way of noting the trend is to examine how frequently the standard was exceeded. For example, Table 2.1 shows the number of SO_2 exceedances of the CEA standard for one-hour average of 0.08ppm – and again we see the dramatic change in the last few months of 2000.

Table 2.1 SO_2 number of daily exceedances/month

	1996	1997	1998	1999	2000
January				3	4
February					5
March			2		5
April		3			1
May		2			
June					
July					
August					
September					
October					2
November			3	1	25
December	5		2	1	40
Total	5	5	7	5	82

Source: NBRO air quality management database

The same general seasonal trends are observed for NOx, shown in Figure 2.2: again, the sharp increase in the 2000 winter peak reflects some new emission source, given the magnitude of the cycles in the previous years.

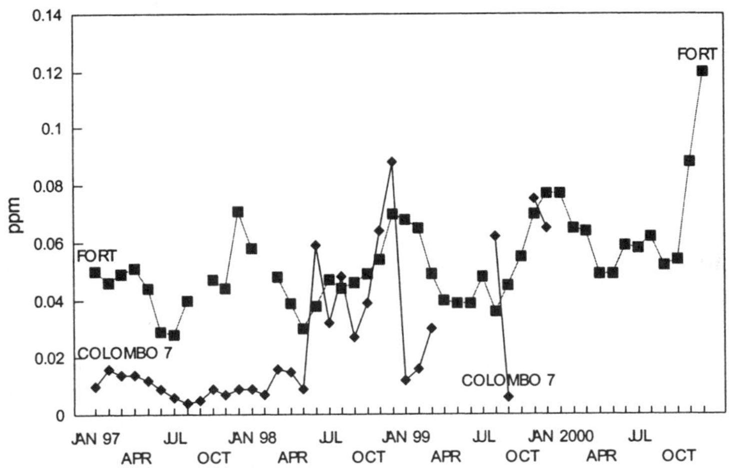

Figure 2.2 Monthly averages for NOx

Figure 2.3 shows the particulate data taken at various sites in Colombo in 1992/93. The seasonalities are less pronounced (except for the Colombo 11 site), but the extent of geographic variation is clear, with average values across sites differing by a factor of 5.

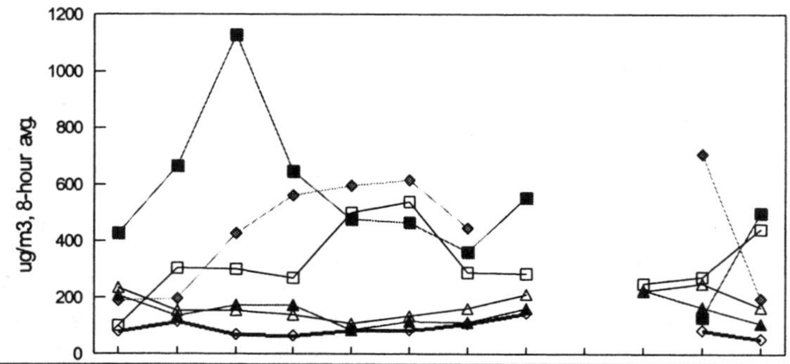

	Sep 92	Oct	Nov	Dec	Jan 93	Feb	Mar	Apr	May	June	July	Aug
■ COL 11	427	665	1128	645	477	467	363	554			132	501
◆ COL 10	190	199	427	561	594	613	447				707	199
▲ COL 5	209	136	173	173	88	115	113	161		229	167	109
⊟ COL 8	101	304	301	269	498	539	289	286		252	274	443
◈ COL 15	84	118	71	66	86	88	109	147			86	57
△ COL 4	235	154	156	141	108	135	162	211		225	250	166

Notes:

COL11	Traffic Police Station, Colombo 11
COL10	Aarasavipaya, Marinade Junction, Colombo 10
COL5	Lanka Tractors Ltd, Nawala Road, Colombo 5
COL8	ICTAD Sub-office, Baseline Road, Colombo 8
COL15	Aitken Spence (Pvt) Ltd, Sri Wickrema Mawatha, Colombo 15
COL4	Milagiriya Church, Colombo 4

Figure 2.3 Averages of 8-hour SPM, ug/m³

There are two problems related to particulate data. First, some fraction of the measured PM-10 is re-suspended road dust – but there appears to be no information on what proportion actually comes from this source. If this accounts for a large proportion, then measures to reduce vehicle emissions will be ineffective in reducing ambient levels.

Second is the question of PM-10 precursors. Some part of ambient PM-10 is the result of secondary sulphate particles whose precursors are SO_2 emissions, and therefore reduction of sulphur content in diesel fuel may affect ambient particulate concentrations. However, in general, the formation of secondary sulphates is a phenomenon of long-range atmospheric transport, involving oxidation over considerable distances, and contributing to ambient PM-10 through dry deposition (and to acid rain during precipitation).

Given the strong sea and land breezes that characterize Colombo's climate (as well as rain on more than 50 per cent of all days), it seems unlikely that dry deposition of sulphates, originating from sulphur emissions of diesel vehicles, would be a significant contributor to ambient PM-10 in Colombo itself. Nevertheless, this is a topic that needs further research.

2.2 THE TRANSPORT SECTOR

Sri Lanka has undergone a series of profound changes in its transport sector over the past 30 years. The most significant was that caused by the liberalization of the economy in 1977, following a long period of strong import controls under socialist governments that severely restricted automobile ownership.

The rapid growth of the automobile fleet (Figure 2.4) illustrates the fundamental changes that occurred after the 1977 liberalization: within five years the number of vehicles per 1000 persons had almost tripled (from around 5.5 before 1978, to 16 by 1983). The Transport Studies Planning Centre (TSPC) estimate for the year 2000 is 60 vehicles per 1000 persons (for Sri Lanka as a whole).

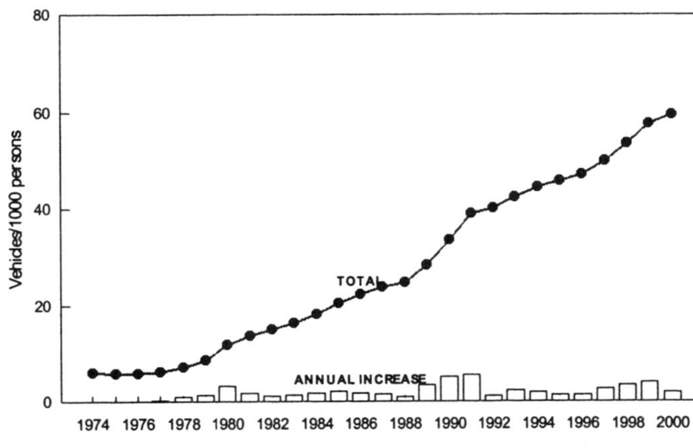

Source: Transport Studies Planning Centre, Ministry of Transport

Figure 2.4 Automobile fleet per 1000 persons (including motorcycles and three-wheelers)

This figure is still well below that of other Asian countries, as discussed later. However, with a third of the total population in the

Colombo metropolitan area accounting for two thirds of the total estimated vehicle trips that occur, the ownership rates in this area would be roughly double the national figure – that is, about 170 per 1000 persons, increasing to 280 per 1000 in Colombo City itself. How much further this vehicle ownership pattern will increase is one of the fundamental issues for Sri Lanka's transport policy.

More recent changes have been no less profound. The past decade of steady economic growth has resulted in sharp changes in the fleet mix. For example, Figure 2.5 highlights the changes in taxi service: imports of three-wheelers from India started only after liberalization in 1978, but have grown rapidly during the 1990s. In the early 1980s the typical Colombo taxi was an aged Morris Minor, replaced after liberalization by reconditioned second-hand Japanese cars: the number of petrol taxis peaked at around 8,700 in 1991, falling to 5,900 by 1997.[5] Clearly, the recent explosive growth of three-wheelers has met a need for urban transportation not met by bus services – but their slow speed, and limited acceleration and manoeuvrability, have severely aggravated traffic conditions in Colombo.

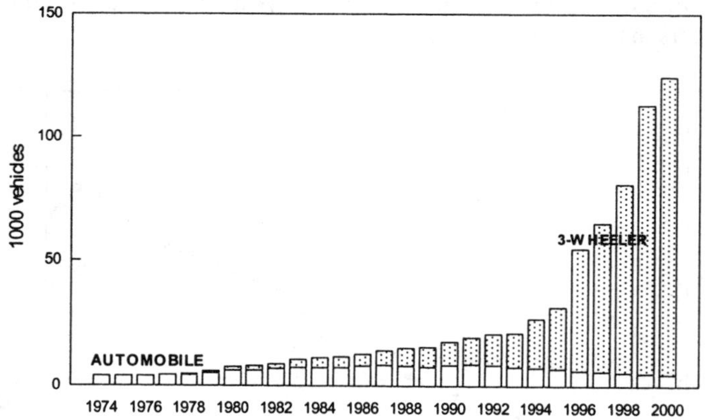

Source: Transport Studies Planning Centre, Ministry of Transport

Figure 2.5 Composition of the taxi fleet

A similar picture is true for motorcycles. Again following liberalization in 1978, and again during 1989–91, the motorcycle fleet increased sharply (Figure 2.6). Though the rate of growth has fallen off over the past few years,[6] we estimate that in the year 2000 there were some 575,000 motorcycles, accounting for 85 per cent of all operational road vehicles in Sri Lanka.

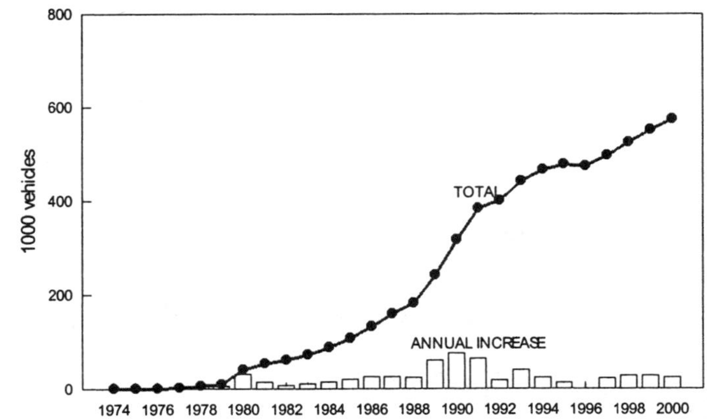

Figure 2.6 The motorcycle fleet

Table 2.2 Person-trips by mode in the Colombo metropolitan area, 1995

	Morning peak		24-hours	
	(000s)	(%)	(000s)	(%)
Rail	68	7.6	226	5.5
Bus	587	65.2	2,601	63.0
Private car or Motorcycle	176	19.5	933	22.5
Taxi, van 3-wheeler	69	7.7	368	8.9
Total	900	100.0	4,128	100.0

Nevertheless, 65 per cent of all motorised trips in the Colombo metropolitan area are made by bus (Table 2.2), and buses account for an even higher share of inter-city travel.[7] As shown in Figure 2.7, there has been an important shift from public to private bus services: most of the growth has been in the private sector (where passenger-km have doubled over the last decade, as opposed to a 30 per cent increase in public buses). However, the quality of data on bus ridership is poor, and these figures are subject to considerable uncertainty.

Rail has seen a decline from 7 to 5 per cent of total passenger-km, despite an increase in the absolute numbers of passenger-km (from 2.7 to

3.1 billion passenger-km per year). Railway electrification – particularly suburban rail services in Colombo – has been discussed for many years. However, the rationale for electrification has changed: in the late 1970s the hope was to reduce oil imports (since at that time the power system enjoyed a surplus of hydro-power); more recently, the main argument for electrification is to reduce road congestion.

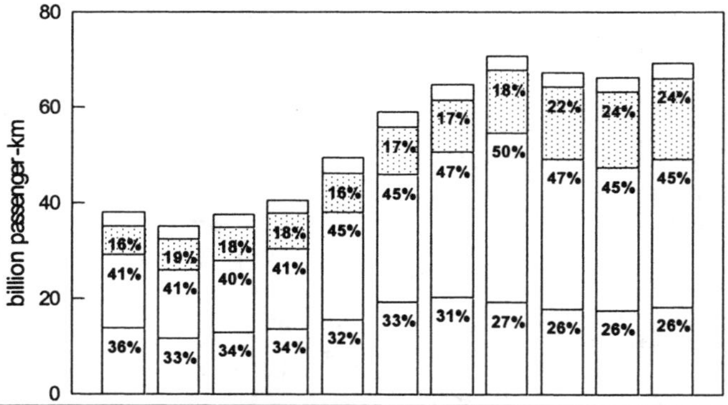

	1990	1991	1992	1993	1994	1995	1996	1997	1998	1999	2000
□ SLCTB	14	11	13	14	16	19	20	19	18	18	18
□ Private bus	15	14	15	17	22	27	30	36	32	30	31
□ Private car	6.1	6.5	6.9	7.5	8.1	9.9	11.0	13.0	14.9	15.8	16.7
□ Railway	2.8	2.7	2.6	2.8	3.2	3.3	3.1	3.1	3.1	3.0	3.2

Source: Transport Studies Planning Centre, Ministry of Transport

Figure 2.7 Passenger-km (modal split, percentage)

The decline in the railway share of freight has been much more pronounced, with the share of tonne-km falling from 18 per cent in 1990 to 1 per cent in 2000 (Figure 2.8). Unlike the passenger rail system (where journey numbers are largely accounted for by Colombo suburban commuter trips, and where absolute numbers have continued to grow), there has been an absolute decline in rail freight. The shift from rail to road freight is a particular problem on the roads leading from Colombo Port, where the growth of container traffic has resulted in large numbers of articulated lorries on the A3, one of Colombo's main radial roads. This has led to the proposal for a so-called 'dry port' located north of Colombo, that is designed to move this traffic back to the rail system.[8]

In comparison to the power sector, where fuel consumption and emissions for each station can be estimated with some confidence, fuel consumption in the transportation sector is known with confidence only in the aggregate –since statistics of Sri Lanka's fuel consumption maintained by the Ceylon Petroleum Corporation are generally reliable. But estimating fuel consumption by vehicle type is a formidable challenge, for there is little actual data, and it is unclear to what extent values published in the international literature are transferable to developing countries.

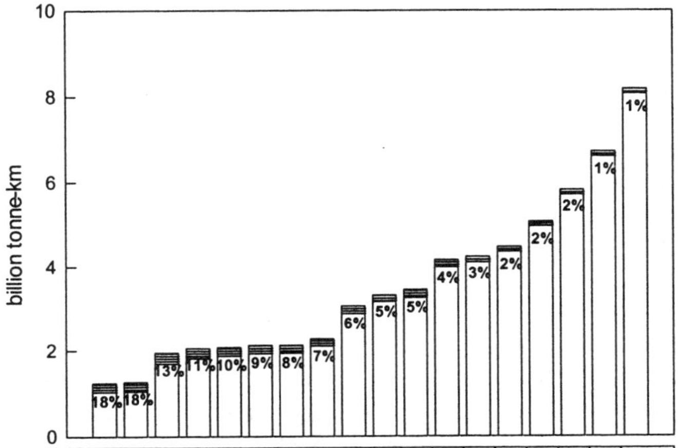

	1983		1985		1987		1989		1991		1993		1995		1997		1999	
□ Road	1.0	1.0	1.7	1.8	1.9	1.9	2.0	2.1	2.9	3.2	3.3	4.0	4.1	4.4	5.0	5.7	6.6	8.1
▤ Rail	0.2	0.2	0.3	0.2	0.2	0.2	0.2	0.2	0.2	0.2	0.2	0.2	0.1	0.1	0.1	0.1	0.1	0.1

Figure 2.8 Freight demand (in million tonne-km)

Motorcycle fuel consumption illustrates the problems. All references in the international literature suggest that two-stroke (2S) motorcycles are less fuel-efficient than four-stroke (4S) motorcycles: the ratios of km/litre for 4S:2S MCs range from 1.30 to 1.65 – that is, 4S MCs are between 30–65 per cent more fuel efficient than 2S MCs. But km/litre estimates themselves vary by a factor of two (from 21–63 km/litre for 4S MCs). Yet this is contradicted by at least some anecdotal evidence in Sri Lanka; at a recent seminar, it was 2S motorcycles that were reported to have a fuel consumption of approaching 60 km/litre.[9]

Now it may well be that a 2S engine of given engine capacity is less efficient (as measured under the controlled conditions of a laboratory test-bed) than a 4S machine – but if, as seems likely, most 2S MCs are of the smallest engine size (50 cc), while most 4S machines are 100 cc (and larger), then the actual average consumption of fuel in the fleet of 2S machines may yet be less per kilometre than in the fleet 4S machines. Unfortunately, few past traffic surveys recorded whether MCs are 2S or 4S, and the available customs data do not record engine size.

In short, reconciling assumptions for numbers of vehicles,[10] fuel consumption per kilometre travelled, and average kilometre travelled per vehicle involves a great deal of judgement, for there is an infinite number of possible combinations of assumptions that all yield the known total annual fuel consumption. In any given year, for vehicles using a given fuel:

$$\sum_j V_j D_j \frac{1}{\phi_j} = Q$$

where V_j = number of vehicles of type j
D_j = kilometre per year in vehicles of type j
ϕ_j = fuel consumption, km/litre, of vehicles of type j
Q = total annual fuel consumption (from CPC statistics)

One of the problems in undertaking this reconciliation is the vehicle classification scheme itself. For example, since taxis are undoubtedly driven significantly more km/year than other passenger cars, a separate accounting for taxis seems desirable – yet there are little good data on taxi fleets. Moreover, it seems likely that taxis represent the group of motor vehicles that has seen the highest rate of conversion to LPG.

Table 2.3 shows our estimate of the 2000 vehicle fleet by type, average yearly kilometre per vehicle,[11] and average fuel consumption.

Table 2.3 Average fuel consumption by vehicle types, 2000

	Fuel	Vehicle fleet	Km/ year	Km/ litre	Fuel use (1000 tonnes)
MC (2S)	Petrol	179,024	5,533	41	18
MC (4S)	Petrol	377,799	5,171	41	35
Car	Petrol	132,929	8,274	10	94
Dual-purpose	Petrol	24,955	5,468	8	15
Land vehicle	Petrol	1,844	4,881	5	1
Three-wheeler	Petrol	73,991	19,676	24	49
Total, petrol		790,542			213
Car	Diesel	20,005	16,759	9	29
Dual-purpose	Diesel	186,488	14,684	8	298
Bus	Diesel	23,948	40,117	3	219
Lorry	Diesel	65,373	22,394	3	351
Land vehicle	Diesel	73,091	3,652	6	36
Three-wheeler	Diesel	3,716	22,112	28	2
Total, diesel		372,621			936

2.3 THE POWER SECTOR[12]

As shown in Figure 2.9, electricity demand growth and GDP growth are highly correlated. Econometric studies conducted by CEB for the 1998 CEB demand forecast shows GDP growth to be the most significant determinant of demand (CEB, 1998a).

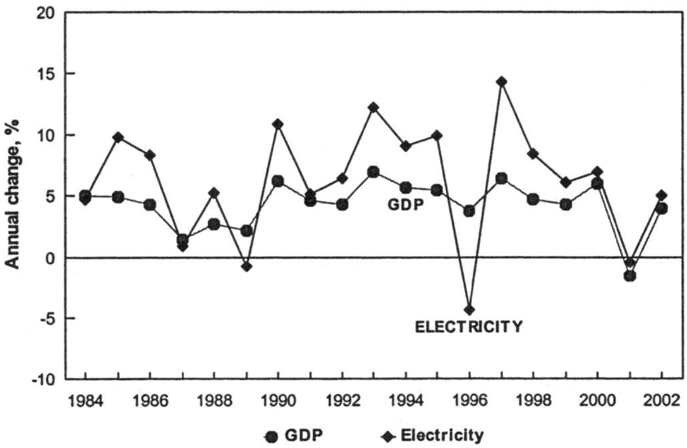

Figure 2.9 Electricity demand growth and GDP growth

Demand has been growing at a compound annual growth rate of around 7 per cent since the early 1970s. However, in 1996 there was an 11 per cent decrease in consumption, largely a consequence of insufficient thermal generating capacity during the drought of that year. The power shortages caused widespread economic dislocation, with a significant loss of economic growth (LIFE, 1998). Yet the most recent load forecasts suggest demand will continue to grow at between 6 per cent and 7 per cent per year, with peak demand increasing from the present 1,285 MW to some 4,300 MW by 2018 (CEB, 1999, Table 2.3). Over the next decade the peak demand is now expected to increase by over 1,500 MW.

Until 1990 thermal generation was used only in dry years (see Figure 2.10), but since then, oil-based thermal generation has been required in rapidly growing amounts. In 1995, 6 per cent of thermal generation was thermal, which jumped to 26 per cent in 1996, and over 50 per cent since 2000. In the dry year 2002, 62 per cent of the total generation was thermal. Lacking any domestic resources, all fossil fuel is imported.[13]

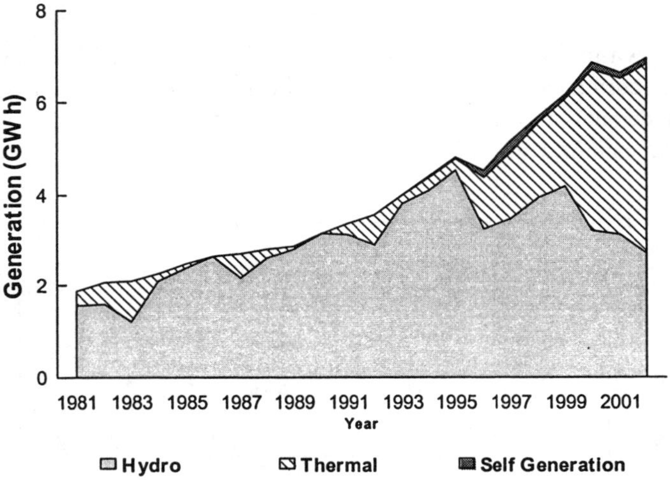

Figure 2.10 Historical power generation mix

Hydro is the main indigenous source of primary commercial energy in Sri Lanka. Of the estimated potential of 2,000 MW, some 1,205 MW has been harnessed to date (Table 2.4).[14] Because of increasing difficulties in implementing large hydro projects, further exploitation of the remaining potential seems unlikely, and most of the capacity additions over the past few years have already been thermal.

Table 2.4 2003 generation and installed capacity, December 2003

	Installed capacity MW	Generation GWh
CEB hydro	1,205	3,188
CEB thermal	560	2,193
CEB wind	3	3.4
IPP thermal	338	1,710
Small hydro	35	126
Hired thermal		390
Total (grid connected)	2,141	7,613

Most of the large hydroelectric sites have already been successfully developed, and the remaining ones have become controversial due to environmental concerns. The larger storage hydro projects also appear uneconomic: the Kukule and Upper Kotmale hydro projects, both run-of-river, may well be the last major hydro projects built in Sri Lanka. And while there are many mini- and micro-hydro projects that could be developed, even under optimistic assumptions about their financial feasibility the total contribution from these sources would contribute no more than 100–200 MW. Table 2.5 shows the major power generation projects of Sri Lanka.

Table 2.5 Installed capacity, MW (as of December 1998)

	MW	Fuel	Commissioning
Laxapana Hydro Complex (CEB)			
Old Laxapana	3 × 8.33 = 25		December 1950
	2 × 12.5 = 25		December 1958
Wimalasurendra	2 × 25 = 50		January 1965
Polpitiya	2 × 37.5 = 75		April 1969
New Laxapana	2 × 50 = 100		Feb–Mar 1974
Canyon	2 × 30 = 60		Mar 1983, Mar 1988
Mahaweli Hydro Complex (CEB)			
Victoria	3 × 70 = 210		1984–86
Kotmale	3 × 67 = 201		1985–88
Randenigala	2 × 61 = 122		July 1986
Ukuwela	2 × 19 = 38		1976
Botawenna	1 × 40 = 40		1981
Rantembe	2 × 24.5 = 49		1990
Other CEB Hydro			
Small hydros	11+6+3=20		1963–88
Samanalawewa	2 × 60 = 120		October 1992
Kukule	2 × 35 = 70		2003
CEB Thermal			
Kelanitissa CT	6 × 20 = 120	Autodiesel	Nov 80–March 82
Sapugaskanda	4 × 18 = 72	Residual oil	May–October 1984
Sapugaskanda: Extension	4 × 10 = 40	Residual oil	Sept 1997, Oct 1999
Kelanitissa CT	115	Autodiesel	August 1997
Combined cycle	105 + 60 = 165	Naphtha	December 2002
Small diesel, Chunnakam	8	Autodiesel	
IPP Thermal			
Lakdanavi	22.5	Fuel oil	1997
Asia Power	50.8	Residual oil	1998
Colombo Power (Barge)	60	Fueloil	June 2000
ACE Power Matara	20	Fuel oil	March 2002
ACE power Horana	20	Fuel oil	December 2002
AES Kelanitissa	165	Autodiesel	August 2003

Source: Ceylon Electricity Board (2002)

The need for baseload thermal generation has been recognized since the mid-1980s when the CEB made its first proposal for a coal-fired project in Trincomalee (Black and Veatch, 1988). However, as noted above, successive delays in implementing the coal project have resulted in oil-fired capacity being added to bridge the supply gap – a strategy that, as we shall see in Chapter 5, will result in significantly higher costs than if the coal project had been implemented as foreseen in the early 1990s. With the acceleration of economic growth since 1990, the decision on how best to meet baseload requirements has reached new urgency.

2.4 THE PETROLEUM SECTOR

In 2003, Sri Lanka consumed about 3.4million tonnes of petroleum products, about half of which (mainly diesel) was imported, the rest produced at the CPC-owned refinery at Sapugaskanda (which imported 2 million tonnes of crude in 2003).

Table 2.6 Petroleum product consumption, 1000 tonnes

	2001	2002	2003
Petrol (90 Octane)	244	277	328
Petrol (95 Octane)	5	9	13
Auto diesel	1,675	1,752	1,652
Super diesel	49	47	42
Kerosene	228	229	212
Furnace Oil	811	780	715
Avtur	138	114(b)	139
Naphtha	14	56	102 (a)
LPG	141	157	161
Total	3,305	3,423	3,360

Notes:
(a) Increase reflects commissioning of the CEB Naphtha-fired CCCT in December 2002
(b) The decline reflects the drop in tourist arrivals following the terrorist attacks at Colombo International Airport in 2001

Source: Central Bank of Sri Lanka (2003)

The Sapugaskanda refinery, commissioned in 1969, is (by international standards) a small, simple hydroskimming refinery. Heavy residual oil is delivered to CEB's diesel plants located adjacent to the refinery. Various proposals have been made to modernize and expand this facility, as it lacks the necessary flexibility (and secondary processing) to match the domestic demands. As discussed in Chapter 7, the growth of auto-diesel (both for power generation and for road vehicles) induced by the distortions in taxation have meant a surplus of naphtha (exported at a loss), and a shortage of diesel (a deficit made up by growing imports). Some argue that the best option is a new, greenfield project (Nexant, 2004) – though given the volatility of refining margins, and the small domestic market, how this facility would be financed remains unclear.

For years there have also been proposals for a new refinery in the South of the country. In 1996 a 200,000 bbl/day refinery was proposed (with an associated 300 MW power project); in December 2003 another proposal was made by Sino Pacific Group Holdings of China for a 165,000 bbl/day refinery with a 700 MW power project. But it is hard to see a small refinery in Sri Lanka becoming a bankable proposition.

NOTES

1 We thank P. Samarakoddy of NBRO for his assistance in compiling the air quality data reported in this chapter.

2 For a complete description of air quality monitoring in Sri Lanka, see, for example, Samarakoddy (1999). Other organizations active in air quality monitoring include the Sri Lanka Atomic Energy Authority (which collected PM-10 and PM2.5 data in 1995–96 for urban Colombo, and 1997–98 in the rural area of Kalutura); and the University of Peradiniya, which collected acid rain samples at 12 locations in 1994–99, and at Nuwara Eliya (in the hill country) since 1996.

3 The various phases of the monitoring programme are described in detail at the NBRO website at http://www.nsf.ac.lk/nbro/enviro.

4 However, since the winter peak is observed in the NOx data (see Figure 2.2), the most likely explanation is some sampling problem for SO_2 during these months, rather than some unusual weather conditions.

5 However, this decrease does not necessarily translate into fewer aggregate vehicle-km by the taxi fleet. A significant share of the Colombo taxi market is now held by radio-cabs, whose utilisation factors are likely to be much greater than previously.

6 Particularly in Colombo, over-capacity has begun to characterize the three-wheeler market.

7 For a history of bus transport in Sri Lanka, see, for example, Diandas (1988); Diandas (1989); or Gomez-Ibanez (1996).

8 See Meier and Munasinghe (2003b) for a detailed discussion of this project.

9 Seminar on Environmental Aspects of Transportation, Institute of Policy Studies, Colombo, March 1999.

10 Estimating the vehicle fleet proves to be an extremely difficult task because of the deficiencies and inconsistencies in the available data, and several past studies have been

based upon estimates that are highly questionable. The fundamental difficulty is that while records are kept to reasonable accuracy on new vehicle registrations, there is no (or very little) data on vehicle retirements or vehicles actually on the road, with the inevitable result of over-estimating the number of vehicles in the fleet. This is compounded by periodic changes in the classification schemes used by official sources.

It is only recently that rigorous research has been conducted on the composition and evolution of the vehicle fleet to the point at which one may have reasonable confidence in models to project the future growth of vehicles and the concomitant growth of emissions, notably Kumarage (1992) and Kumarage and Wijertane (1999). This work involves careful reconciliation of the available data sources; the formulation of appropriate models of vehicle growth based on hypotheses about survival rates: the calibration of model results with the empirical data (as obtained, for example, from traffic counts), and reconciliation of estimated vehicle fleets, fuel efficiency, and typical annual distances travelled by each class of vehicle with the total consumption of fuel (known from data published by the Ceylon Petroleum Corporation).

The methodology for this fleet estimation process, as further developed for the purposes of this book, is given in the detailed transportation sector report (Meier and Munasinghe, 2003b).

11 The data for average km/vehicle shown here are substantially higher than those reported in the early 1990s, and which were used as a basis for the Metropolitan Environmental Improvement Plan (MEIP). For example, for motorcycles, MEIP assumed 7,451 km/year, as opposed to 5,100–5,500 km used here. For heavy lorries, MEIP assumed 38,000 km/yr, as opposed to our estimate of 22,400 km/yr; and for cars, MEIP assumed 15,000 km, as opposed to our estimate of 8,275 for gasoline and 16,750 km for diesel cars. The estimates by Walsh (1992) are even higher, e.g. 50,000 km/year for passenger cars, and 48,000 km/year for lorries.

The best evidence that these estimates (together with assumed vehicle fleet and fuel efficiency assumptions) are far too high is that the resulting fuel consumption does not reconcile with the figures of total gasoline and diesel consumption (known with a high degree of accuracy from CPC data). This is unfortunate, for it resulted in estimates of emissions that were also far too high.

For example, from refinery data we know that the average addition of TEL to gasoline was 0.23 gm/litre (significantly less than the specification value of 0.45 gm/litre), which when applied to the known quantity of gasoline delivered, results in a total addition of 53 tonnes. However, Walsh (1992) estimated lead emissions as 233 tonnes, apparently using the specification rather than the actual rate of TEL addition. Even if this estimate is scaled by the ratio of actual to specification TEL addition, the resulting value of 119 tonnes (i.e. 233 x 0.23 x 0.45) is still double the actual rate based on known gasoline consumption.

12 For a more detailed presentation of Sri Lanka's power sector, and its historical development, see for example, Meier and Munasinghe (1994). A detailed presentation of Sri Lanka's energy resources is to be found in Baldwin (1991).

13 There are some small peat resources in marshy lands to the North of Colombo, but a 1989 feasibility study concluded that the quality and extent of the reserve was not commercially viable.

14 Including the 70 MW Kukule plant.

3. Valuation of health damage costs

We begin with a review of health damage studies in Sri Lanka – of which there are few. We then make estimates of the health damages from local air emissions using the benefit transfer method. We also briefly review the health effects of leaded gasoline and make a preliminary estimate of an appropriate value for Sri Lanka.

3.1 HEALTH EFFECTS

There exists to our knowledge only one Sri Lanka-specific study of health damage valuation related expressly to vehicular emissions (Chandrasiri and Jayasinghe, 1998). Although there are many limitations (as noted by the authors), this is also the only study that is based on clinical data.

Table 3.1 Vehicular particulate emissions and hospital admissions/deaths from respiratory diseases

	Particulate emissions	Hospital admissions(a)	Hospital death rate(b)
1991	435.3	1,435	1,217
1992	504.7	1,326	1,326
1993	612.4	1,609	1,435
1994	746.4	1,539	1,379
1995	793.4	1,561	1,488
1996	731.9	1,626	1,657

Notes:
(a) Admissions for diseases of the respiratory system (excluding URT diseases, pneumonia, broncho-pneumonia and influenza), per 100,000 population
(b) Hospital deaths from the same causes, per 1000 population

Source: Chandrasiri and Jayasinghe (1998)

As shown in Table 3.1 (and Figure 3.1), over the past decade there has been a relentless increase in hospital admissions and deaths related to respiratory disease,[1] that tracks very closely the increase in vehicle emissions of particulate matter.[2]

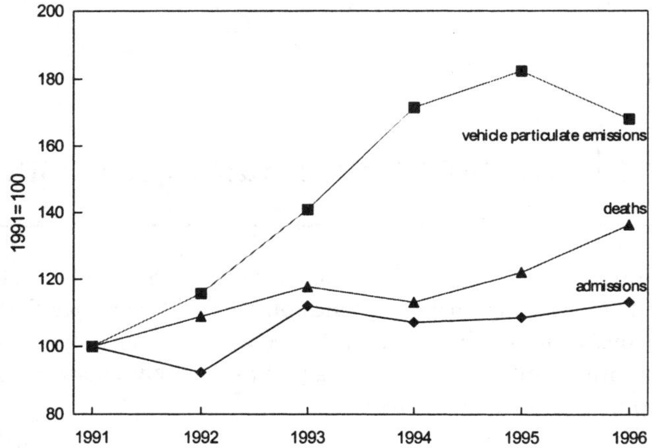

Source: Chandrasiri and Jayasinghe (1998), Table 16

Figure 3.1 Vehicular particulate emissions and hospital admissions/deaths from respiratory diseases

The overwhelming majority of the hospital physicians interviewed asserted that the increase in respiratory diseases was related to vehicle emissions.[3] This was confirmed by a study of individuals exposed to high emissions levels – trishaw drivers, traffic wardens and traffic police officers. Controls were matched for age, sex and social class. Each interviewee was asked about the incidence of symptoms related to respiratory problems experienced over the past month, and since starting their occupation (sore throat, wheezing, coughing, shortness of breath, headache and eye pain/irritation, etc.).[4] The results (Table 3.2) show clearly that the extent of such complaints in these occupations is significantly higher than those of the control groups.

There are a number of potential objections to this study (which are again noted by the authors).[5] Most importantly, notwithstanding the correlation in the data, causation is far from being established, and individuals' judgements as to what is the cause of their health problems require great care.[6] However, when combined with the hospital admissions and death rate statistics, and additional surveys of patients visiting general practitioners working in Colombo, the study suggested at least some linkages beyond mere correlation.

Table 3.2 Incidence of symptoms/diseases during previous month

	Trishaw drivers (*n* = 50)		Traffic Police Officers (*n* = 25)		Traffic Wardens (*n* = 25)	
	Test	Control	Test	Control	Test	Control
Sore throat	6 (12%)	0	2 (8%)	0	2 (8%)	0
Fever	3 (6%)	1 (2%)	3 (12%)	1 (4%)	2 (8%)	1 (4%)
Asthma	0	0	0	0	0	0
Chest pain	0	0	1 (4%)	0	0	0
Shortness of breath	0	0	2 (8%)	0	1 (4%)	0
Cough	11 (22%)	0	13 (52%)	0	3 (12%)	0
Sneezing	1 (2%)	0	1 (4%)	0	0	0
Backache	0	0	7 (28%)	0	1 (4%)	0
Upper limb Pain	2 (4%)	0	3 (12%)	0	0	0
Lower limb Pain	0	0	5 (20%)	0	0	0
Abdominal Pain	2 (4%)	0	1 (4%)	0	2 (8%)	0
Dizziness	0	0	1 (4%)	0	0	0
Headache	7 (14%)	0	5 (20%)	0	1 (4%)	0
Eye	6 (12%)	0	1 (4%)	0	2 (8%)	0

Source: Chandrasiri and Jayasinghe (1998), Appendix 5

Senanayake et al. (1999) analysed hospital admissions of children for asthma attacks, and found significant correlation between the highest daily rates of wheezing requiring immediate nebulization, and the highest daily maximum one-hour averages of SO_2 and NOx (and similar correlation for the minimum rates). However, the relationship between asthma and air pollution is controversial: the UK Committee on the Medical Effects of Air Pollution (1998) notes that the evidence does not support a causative role for outdoor air pollution, and opines that pollen and fungal spores are more likely to be influential.[7]

3.2 LOCAL AIR EMISSION DAMAGE VALUATIONS

Ideally, environmental damage costs should be measured directly (see for example, Munasinghe 1992). In this study we use the benefit-transfer method of damage valuation – the basis of which is the extrapolation to Sri Lanka of valuations estimated elsewhere. While there is little doubt that such an approach has its shortcomings,[8] and that country specific damage valuations would be far preferable, where the latter do not exist there is little choice but to use the former.[9] Indeed, the approach has been used in a number of recent developing country studies:[10] provided the limitations are recognized, the method can provide useful indicative valuations of likely damage valuations.[11] It has also been used in Sri Lanka – in a study of the impacts of the 1996 power shortage, (LIFE, 1998), and in the Environmental Impact Assessment of the Colombo–Katunayake Expressway, (RDA, 1997).[12]

The method as presented by the Asian Development Bank (ADB) (Abeygunawardena et al., 1999) requires the following steps:

1. Specify the base year unit damage values (as US$/tonne per 1000 persons). For this purpose ADB uses the study by Rowe et al. (1995) for New York State. The corresponding values are shown in Table 3.3
2. Adjust the price levels to the year 2000 using US GDP growth.
3. Multiply the values by the ratio of per capita GDP. In 2000, the per capita GDP in Sri Lanka, adjusted for purchase power parity (PPP),[13] is US$3,900; that of the US is US$34,000.
4. Multiply the values by the estimated population resident within each radius.
5. For future years, adjust by the estimated population growth rate (in each radius) and GDP growth in each radius.

Table 3.3 Damage estimates, 1992, US$/per tonne/per 1000 persons

	Local (within 30 km)		Regional (within 30–80 km)		Distant (80–500 km)	
	Low	High	Low	High	Low	High
Particulates	0.20	0.30	0.10	0.30	0.04	0.08
SO_2	0.04	0.08	0.02	0.04	0.01	0.02
NOx	0.06	0.08	0.04	0.06	0.01	0.02

Application of this approach yields the values shown in Table 3.4.[14]

Table 3.4 Damage estimates, ADB benefit transfer method, US$/tonne

	Low	High
Sri Lanka 2000		
Particulates	258	478
SO_2	53	106
NOx	81	116
Sri Lanka 2020		
Particulates	1,623	2,779
SO_2	327	654
NOx	506	697

The valuations derived by LIFE (1998) in the evaluation of the impacts of the 1996 power crisis are shown in Table 3.5.[15] However, based on more recent studies (such as the Six Cities Study, discussed below), discussed below, we believe these underestimate the damages from vehicular sources.

Table 3.5 Damage valuations (US$/tonne)

	Source Document	Adjusted to 2000 using PPP
NOx	51.8	334
TSP	75	483
SO_2	128	825

Source: (LIFE, 1998), Annex V

A study by the University of Moratuwa (Kumarage, 1999) used these values to estimate damages in terms of Rs/litre for different vehicle types.[16] As indicated in Table 3.6, these show very little variation across vehicle types when expressed in Rs/litre: but when converted to Rs/km travelled, the environmental damages show large variation, as one might expect.

Chandrasiri and Jayasinghe (1998) estimated the value of health damages attributable to particulates in Colombo at between Rs 67–160 million/year, based on limited data of highly exposed occupational groups (traffic wardens, trishaw drivers, traffic policemen). A World Bank assessment assessed the health damages in Colombo attributable to PM-10 at US$ 30 million, based on an average ambient concentration of 40–50 $\mu g/m^3$ (World Bank, 2000). However, it is unclear upon what basis either the average exposure, or the unit health damage, was derived.

Table 3.6 University of Moratuwa damage valuations

	Rs/litre(a)	Rs/vehicle-km(b)
Car	0.57	0.06
Motorcycle	0.56	0.01
Three-wheeler	0.62	0.02
Van	0.65	0.07
Medium bus	0.71	0.14
Large bus	0.79	0.20
Medium lorry	0.71	0.14
Large lorry	0.79	0.23

Notes:
(a) Kumarage (1999), Table B4-2
(b) Based on average fuel efficiencies given in Table 2.3

The Six Cities Study

Another recent source of damage valuations is the 'Six Cities Study' (Lvovsky et al., 2000). This study permits the benefit transfer method for estimating health effects to be applied to vehicular emissions. Since the majority of health effect valuation studies in the US and Europe relate to power sector emissions (typically from tall stacks in relatively remote locations), there are problems in applying the benefit transfer method to vehicular emissions – emitted at ground level in generally densely populated areas. Based on studies of six cities (Mumbai, Shanghai, Bangkok, Krakow, Santiago and Manila), the report derives representative damage cost estimates for vehicular emissions of PM-10, SO_2 and NOx (Table 3.7) – and which for the four Asian cities are 30–50 times greater per tonne than for power plant emissions from high stacks (and even greater in the case of Krakow and Santiago).

Table 3.7 Marginal damage costs (US$/tonne)

		Power plant (high stack) [1]	Large industry (medium stack) [2]	Small boilers and vehicular emissions [3]	Ratio vehicular: power plant [4] = [3]/[1]
Munbai	PM-10	234	1,077	7,963	34
	SO_2	51	236	1,747	34
	NOx	20	93	668	33
Shanghai	PM-10	161	502	5,828	36
	SO_2	36	112	1,295	36
	NOx	11	33	385	35
Manila	PM-10	345	1,828	17,942	52
	SO_2	61	324	3,183	52
	NOx	24	129	1,265	52
Bangkok	PM-10	828	2,357	28,722	34
	SO_2	147	417	5,087	34
	NOx	57	162	1,971	34
Krakow	PM-10	97	682	13,255	136
	SO_2	18	130	2,522	140
	NOx	4	29	560	140
Santiago	PM-10	692	4,783	88,551	128
	SO_2	132	911	16,864	127
	NOx	35	240	4,445	127

Source: Lvovsky et al., (2000) Table 1.5 (p.16)

One must recognize the large uncertainties in damage cost estimates, and most studies show considerable variation. Table 3.8, taken from a study of renewable energy in Croatia, shows different estimates of local damage costs from fossil pollutants revealed in European studies: we note that the values of the Six Cities study are at the low end of the range of values (especially when compared to the 'Extern E' estimates shown in Table 3.8, which are generally recognized as being very high). Uncertainty in damage cost estimates are of particular concern for studies that seek to determine the economically optimal level of renewable energy (see Box 3.1).

Table 3.8 European coal generation damage cost estimates

Study	SO_2 (€/tonne)	Particulates (€/tonne)	NOx (€/tonne)	Total damage cost for coal (€cent/kWh)
Croatia, Zagreb (EKONERG, 2001)	13,483	24,218	19,265	6.0
UK, ExternE (AEA, 1998)	6,820	14,060	5,740	2.0
'Representative European values' ExternE (Spandaro and Rabl, 1998)	10,450	15,400	15,700	4.8
EU DG Environment (BeTa Database) (Holland and Watkiss, 2001)	5,200	14,000	4,200	1.6
Portugal, ExternE (CEETA, 1998)	4,959	5,975	5,565	1.8
World Bank, Six Cities Study Lvovsky et al., 2000)	96	1,723	255	0.1

Source: Frontier Economics (2003)

Mindful of the cautions emphasized by the authors of the Six Cities Study, the suggested indicative marginal unit values shown in Table 3.9 still provide useful guidance.

Table 3.9 Unit health damages, US$/tonne per 1,000,000 population per US$1000 per capita income

One tonne change in emission of:	High stack (power plants)	Medium Stack (large industry)	Low stack (boilers and vehicles)
PM-10			
Range for six cities	20–54	63–348	736–6,435
Average	42	214	3,114
SO_2			
Range for six cities	3–8	10–56	121–1,037
Average	6	33	487
NOx			
Range for six cities	1–3	3–13	29–236
Average	2	9	123

Source: Lvovsky et al. (2000), Table 6-3, (p.65)

Box 3.1: The economically optimal level of renewable energy

Decisions about implementing renewable energy targets are pending in several countries as mandated market share mechanisms are being considered. The standard methodology to estimate the economically optimal quantity of renewable energy is to derive a supply curve for renewable energy, and then set the renewables target at the quantity at which the costs of the marginal plant are exactly equal to the avoided costs (including the avoided environmental damage costs).

The figure below illustrates the idea. We assume that the cost of fossil fuel generation (c), the externality value (v_{ENV}), and the supply curve (S) are all known with certainty. The optimum quantity of renewable energy is given by the intersection of the supply curve S with the avoided economic cost (including the value of the externality cost), that is, by the point α. It follows that the economically optimal target is Q_{IN}, which if implemented by a system of tradable green certificates (TGC), results in a TGC price equal to v_{ENV}.

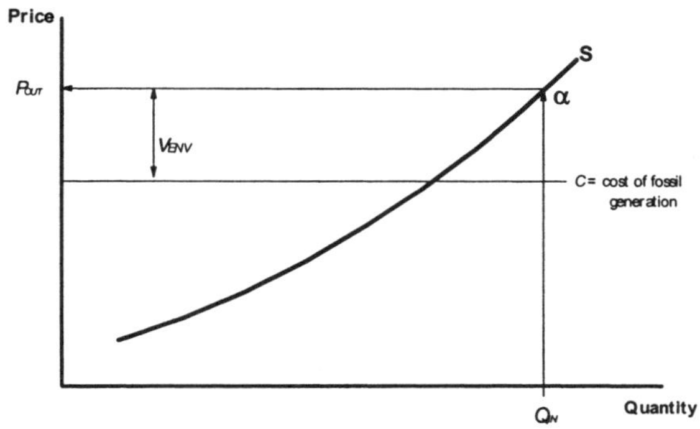

Clearly if the actual damage cost were twice the assumed value of v_{ENV}, then a target set at Q_{IN} would significantly underestimate the true economic optimum (and overestimate the optimum if the true damage cost were lower than the assumed value). This in turn has implications for the policy instrument used: if all quantities are known with certainty, then it matters little whether government sets a quantity target at Q_{IN} (as in many American States or Australia), or sets a price at P_{OUT} (as in Germany, where it is known as the Feedlaw), for the result will be the same. But in the face of uncertainty, which may require a change in policy if new research findings reveal significantly higher (or lower) damage costs, experience shows that adjusting a quantity target is likely to be much easier than adjusting a price (as vested interests defend the status quo). For further discussion, see Meier (2004).

For example, if we apply the lowest of the values in the ranges given to Colombo (with a population of 4.9 million, and GDP/capita of US$ 4,933), then the values compute to US$ 17,790/tonne for PM-10, US$ 2,924/tonne for SO_2, and US$ 700/tonne for NOx, values comparable to those for Manila. Table 3.10 summarizes the various estimates, converted to Colombo for 2000.[17]

Table 3.10 Comparison of damage valuation estimates (in US$/tonne)

Basis	Source	Colombo 2000
Particulates (a)		
Six Cities Study, low value of range	PM-10	17,790
	as TSP	14,232
ADB benefit transfer method	Table 3.4	258–478
RDA, CKE environmental assessment	See text	8,500
LIFE	Table 3.5	483
University of Moratuwa		262
SO_2		
Six Cities Study, low value of range	See text	2,924
ADB benefit transfer method	Table 3.4	53–106
LIFE	Table 3.5	825
University of Moratuwa		447
Nox		
Six Cities Study, low value of range	See text	700
ADB benefit transfer method	Table 3.4	506–697
LIFE	Table 3.5	334
University of Moratuwa		180

Note:
(a) Only the Six Cities study gives PM-10 health damage costs; the others are for total particulate matter. In the case of ambient air quality, the Six Cities study uses a PM-10/TSP ratio of 0.55; however emissions of PM-10 are estimated, on average, to be 0.8 of the TSP emission factor for diesel vehicles, and 0.9 of the TSP emission factor for gasoline vehicles (Lvovsky et al., 2000, Annex A). The proportion of PM-10 in total TSP from typical power plant stacks is smaller, about 0.6.

We recognize that limiting the monetization to PM-10, SO_2 and NOx underestimates the total health damage. A study in the UK (Glaister et al., 1999) provides some estimate of the relative damage costs across emissions (Table 3.11): of the total damage cost of 2.424 pence/km for urban diesel emissions, 70 per cent is attributable to PM-10 alone, and 82 per cent to the sum of PM-10, SO_2 and NOx.[18]

Table 3.11 Health damage in London (diesel vehicles)

	Pence/km	Percentage of total
Particulates	1.692	70
SO_2	0.182	8
NOx	0.113	5
Sub-total	1.987	82
Sulphate aerosol	0.027	1
Nitrate aerosol	0.219	9
Ozone from NOx	0.098	4
Ozone from VOC	0.041	2
Benzene	0.052	2
Total, all emissions	2.424	100

Source: Glaister et al. (1999), p. 54

3.3 LEAD

There are no systematic health damage studies of lead in Sri Lanka.[19] However, there is a vast international literature on the subject, too extensive to review here.[20] In the US, the benefit–cost ratio of eliminating lead in gasoline was estimated as being in excess of 10 (USEPA, 1985); one estimate of the health benefits shows that reducing the US populations' blood level by 1 μg/dl was US$ 18.9 billion per year.[21] A recent World Bank review of this literature notes that 'the magnitude of these benefits suggests that phasing out lead from gasoline is likely to produce substantial benefits in other countries as well' (Lovei, 1998) – a conclusion evidently shared by most governments in Asia that have phased out leaded gasoline.

Table 3.12 shows the results of the one study in Sri Lanka that measured lead levels in blood (Awegoda et al., 1994) with comparisons from elsewhere (though these need to be seen with a great deal of caution, given differences in sample populations, age groups, year of sampling on so on). Nevertheless, despite these caveats, the high levels of lead in the most exposed groups compared to the control population are evident,[22] although the general levels do appear to be less than half of those in other large Asian cities.[23]

Table 3.12 Average lead levels in blood

Sri Lanka (Colombo)	mean, µg/dl
School children (4–5 years)	5
Motorcycle riders	12
Street vendors	13
Three-wheeler drivers	15
Traffic police	53
Adult control group	9
Other locations	
Bangkok	40
Manila	20
Jakarta	15

Sources: International comparisons from Lovei (1998, Figure 3); Sri Lanka data from Awegoda *et al.* (1994).

Mindful of the hazards of extrapolation across countries, we may nevertheless make an order of magnitude estimate for the corresponding health damage in Sri Lanka. If we scale the estimate by ratio of population and ratio of PPP adjusted GNP/capita, then a reduction of average concentrations of lead in blood of 1 µg/dl computes to a benefit of Rs 19 billion (for the year 2000), about 1.5 per cent of GDP.[24] Given year 2000 gasoline consumption of 277 million litres/year, this in turn translates to Rs 68.7/litre of gasoline. Even if this estimate were in error by an order of magnitude, (for example, under the even more conservative assumption that elimination of all gasoline lead emissions would reduce blood lead by only 1 µg/dl),[25] the benefits of unleaded gasoline would still correspond to around Rs 7/litre. Notwithstanding the uncertainties of this calculation, it does seem to confirm the conclusion of the World Bank that phasing out lead is likely to produce substantial benefits.

NOTES

1 These data are based on countrywide hospital admissions as recorded by the Annual Health Bulletins published by the Ministry of Health. The in-patient data from private sector hospitals were not available, but these account for only a very small proportion of the total (less than 5 per cent).

2 Since the bulk of the increase in vehicle emissions is in Colombo, but the health statistics are for the country as a whole, one may expect (as suggested by Figure 3.1) the former to increase faster than the latter. There is also very likely a lag effect.

3 Of general practitioners (GPs) interviewed, 94 per cent noted sharply increased visits for problems of the upper respiratory tract over the past few years, not explainable by other

specific causes (Chandrasiri and Jayasinghe, 1998; Table 13). However, there remains the question of the extent to which these opinions adequately take into account the actual incidence of smoking. We have no data for Sri Lanka, but smoking rates have increased sharply in many other Asian countries.

4 Information about smoking and alcohol consumption was available only for trishaw drivers – among whom the incidence of smoking (44 per cent) was lower than the control group (68 per cent), so smoking does not explain the much higher rates of complaint.

5 This was a small sample size with no information on smoking or alcohol consumption for the traffic policemen, etc.

6 In a recent opinion poll in London, 60 per cent of respondents cited poor air quality as having the most effect on their health – compared to 27 per cent who cited dirty streets (litter, dog droppings); 26 per cent who cited smoking, and 14 per cent who cited insufficient exercise (Glaister et al., 1999).

7 Indeed, in London, between 1976 and 1997, while asthma increased by 300 per cent, emissions of air pollutants over the same period decreased by 180 per cent (Glaister et al., 1999).

8 Freeman (2000) lists the main issues as follows: 1. if pollution levels are higher in the developing country than is the country that is the source of the dose-response function, then the transfer involves extrapolation that is outside the range of the data; 2. the evidence is that the elderly are more sensitive to the life-shortening of particulate matter . . . therefore in developing countries with a younger age structure there may be an overestimate on premature mortality; 3. the amount of time spent outdoors may be different (and lead to an underestimate in tropical countries where individuals spend more of their time outside the home).

9 The potential issues involved in the benefit transfer method have been extensively discussed elsewhere (see for example, World Bank, 1994, in the case of the Bangkok study). The consensus view in the literature is that it is likely that US-based estimates of the dose-response functions are conservative when applied to developing countries. Given that most individuals in Colombo spend a much higher proportion of the day on streets than do most Americans, and that ventilation in urban residences in Colombo is much greater than in air conditioned US houses (or tightly insulated US houses in winter), it is likely that a Colombo resident's exposure to ambient (outdoor) concentrations is much greater (and that exposure to indoor sources is lower) than that which determines dose-response in US epidemiological studies.

The Bangkok study also notes other reasons why US-based dose-response functions are likely to be conservative, including that the average health status in a developing country is likely to be lower (and therefore individuals are more susceptible to pollution-induced diseases), and that lower incomes limit the more costly avoidance options available in the USA.

10 For example, several recent studies of energy sector policies and projects in China have used this approach. For application to an assessment of renewable energy generation, see Meier (2004); for an application to a power transmission project, see Abeygunawardena et al., (1999).

11 Thus we are mindful of the admonitions of Freeman (2000): ' . . . analysts should be careful to discuss the issues raised here, properly qualify their results, and be honest about the uncertainties introduced by this practice [of the benefit transfer method].'

12 We note some significant errors in the application of the method in the CKE-EIS, which used UK damage values as a basis, and which are documented in detail in Meier and Munasinghe (2003b): some of the damage cost estimates are overstated by a factor of two.

13 In this report we use a PPP adjustment factor of 4.56 (based on the data tables in the World Resources 1998–99, which shows 1995 per capita GDP of US$ 720 at the official exchange rate, US$ 3,290 at PPP).

14 There has not been a comprehensive population census for two decades, and official figures are based on small household surveys which provide little or no information on migration at the district level. The official population annual growth figures for 1998/99 are 1.26 per cent for Colombo, 1.4 per cent for the region, and 1.5 per cent for the balance of the Island, or 1.4 per cent nationally. However, these figures are based only on birth and death rates, and do not take into account migration, which accounts for a significant proportion of the growth in Colombo, and for which an additional 2.5 per cent per year may well be quite conservative.

Again comparisons with the 1994 study of congestion costs in Bangkok (World Bank, 1994) are instructive: annual population growth in Bangkok was estimated at 2.2 per cent, compared to 1.8 per cent for Thailand as a whole (though as in the case of Sri Lankan data, it is unclear whether these official estimates account for natural growth only, or whether they include net migration from rural areas).

15 These estimates are based on the classic study by Pace University (1990), and reflect average values for the United States. The Pace estimates suffer from the deficiency that they do not expressly take into account population densities, and, obviously, are applicable primarily to power sector emissions.

16 Source document values as given in Table 3.5, but updated to 1999 rather than 2000 (and without adjustment for PPP): US$ 262/tonne for particulates, US$ 447/tonne for SO_2 and US$ 180/tonne for NOx; and US$ 15/tonne for CO; Pace (1990, Table B4-1).

17 Using the corrected values for the CKE study, based on the UK value of 4.5Rs/gram, translates to US$ 56,250/tonne. Unfortunately, no source is given in the CKE-EIS for this UK data point, which also seems unusually high (compared to most other estimates in the literature).

18 Since unleaded petrol is no longer sold in London, lead does not appear in this table.

19 Neither Chandrasiri and Jayasinghe (1998) nor Senanayake et al.. (1999) included discussion of lead in gasoline (though the former does contain the data on ambient lead measurements taken in 1992/93).

20 For a good review, see, for example, Lovei (1998).

21 Schwartz (1994). Given a 1989 GDP of around US$ 6,800 billion, this represents 0.25 per cent of GDP (which is a very high fraction, given that this represents the damage for just one pollutant!).

22 Since these measurements were made, an additional group of highly exposed individuals may be noted, particularly in Colombo, namely the large numbers of police and army personnel on security duty, largely a roadside occupation. While it may be hoped that this is not a permanent feature of Colombo life, and while one may argue as to whether the incremental (occupational) health risk (compared to the much higher risk of death or injury from terrorists) qualifies as an externality of the transportation sector, we would expect that the long-term health effects of these individuals' lead exposure are no different to that of the occupational groups that were included in the study.

23 According to the Clean Air 2000 Action plan, average ambient levels of lead in 1992/93 were assessed at 0.38 × 10-3 mg/m^3 (8-hour average) during the dry weather periods (MEIP, 1994, p.9).

24 That is, US$ 17.152 million × 34,000/856*4.56 (PPP adjustment factor) × 250/19.4 (population ratio).

25 So high an externality value might well raise some doubts. But even higher orders of magnitude have been encountered in other countries: for example, in China, total environmental damages of air and water pollution were estimated at 8 per cent of 1995 GDP (see World Bank, 1997c).

4. Economics of imported fuels

An understanding of the behaviour of fuel costs, and the differentials between them, is fundamental to an appreciation of the economics of technology choices. Indeed, as one reviews the various CEB planning studies, detailed feasibility reports, and proposals advanced by advocates of particular fuels and technologies, it is evident that fuel price forecasts have often lacked consistency and fail to state clearly the underlying assumptions. In a period of unprecedented volatility of international oil prices, achieving consistency may well be difficult: however, the importance of clarity of assumptions and consistency of economic and financial comparisons cannot be over-emphasized.

4.1 PRICE CONSISTENCY

We begin with coal. Australian and South African export coals are the major potential sources for Sri Lanka, though over the last few years Indonesian coals have also become more available.[1] Although coal prices do not exhibit the short term volatility of world oil prices, the longer-term averages show remarkable correlation to the annual average oil price. This is shown in Figure 4.1, where a comparison is made of annual average Brent and coal (fob New South Wales) prices from 1981 to 2004.[2]

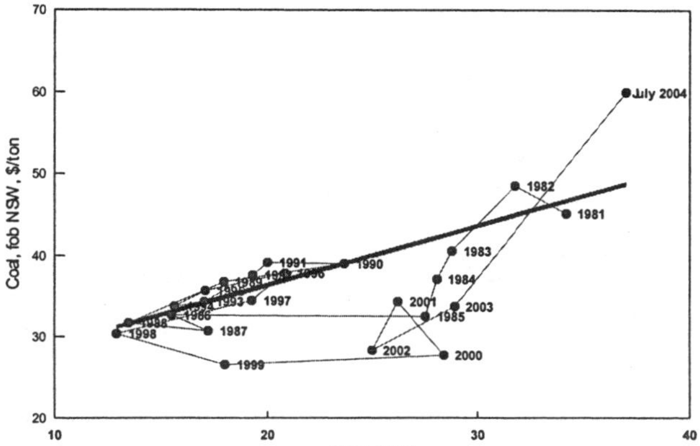

Figure 4.1 Annual average coal and crude oil prices

As is evident from Figure 4.1 there have been two periods over the last 20 years when the general relationship between coal and oil prices has shown significant deviations. The first was in the years 1983–85, prior to the oil price collapse of 1986 – this period was characterized by an artificially high crude-oil price as OPEC tried to defend its posted pricing system for crude oil, despite falling product prices. This effort was not sustainable, leading to the spectacular fall in crude oil prices in late 1986.

The second period of price aberration has been the past few years – with coal prices (in the Asia-Pacific markets) significantly below their normal values. This was a consequence of the aftermath of the financial crisis of 1997/98, and the rapid growth of China's coal exports, which both depressed prices. Low coal prices in the Asia-Pacific export coal market prevailed until 2003, but in the first seven months of 2004, Australian spot coal prices have almost doubled. However, one may expect that if Brent reverts to a trading range of US$ 30–40/bbl, then coal prices at US$ 60/tonne are not sustainable, and should also revert to their long-term relationship with oil. Since heavy fuel oil is the main alternative to coal for power generation (in baseload, steam-cycle plants), coal cannot deviate from the equivalent heavy fuel oil price for much more than a few years (though much of the sharp recent rise in coal cif prices is attributable to freight shortages as coal demands in the region have picked up).

Most useful for economic analysis is to express prices in terms of heat value, for which purpose the set of units that is most widely used in energy economics is million BTU. As is evident from Figure 4.2, the coal price is much less volatile.

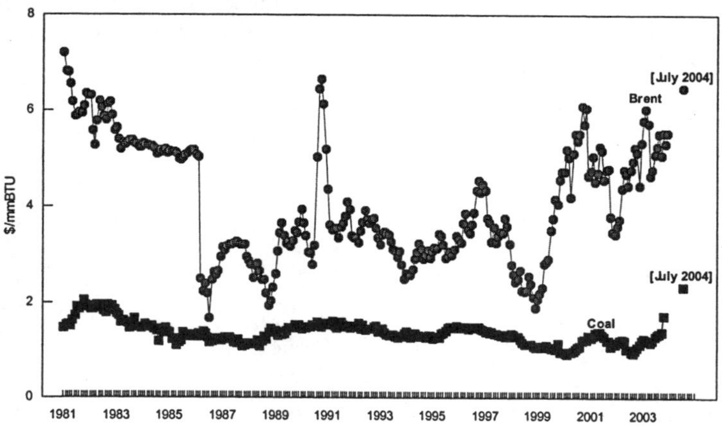

Figure 4.2 Brent v. Australian coal as US$/millionBTU, fob

Forecasts of world energy prices are generally for one of the so-called "marker" crudes (such as Brent, West Texas Intermediate (WTI) or Dubai/Oman),[3] or for Saudi (Arab) Light. Following the detailed analysis of Asian oil markets by Horsnell (1997), in this study we take Brent crude as the marker and express prices of all of the relevant petroleum products as fractions of the Brent price.[4] Product prices on spot markets tend to be even more volatile than crude prices, and are often accentuated by market perceptions of inventory (in)adequacy; during the 1990/91 Gulf Crisis, while crude prices reached a peak in October 1990 and then dropped to US$ 23/bbl by January 1990, diesel prices had a second peak in February 1991 (reaching US$ 42.50/bbl).

The two petroleum products that are of most interest for power generation in Sri Lanka are auto-diesel (used in combined cycle gas turbines as well as the smaller, diesel IPPs), and heavy fuel oil (as may be used in larger diesels or steam cycle oil plants). The relevant prices are as given by the Singapore (or Gulf) spot markets, plus freight to Sri Lanka; similarly, the coal price needs adjustment for transportation from Australia. These cif prices are shown in Figure 4.3, in US$/million BTU, but adjusted to the west coast of Sri Lanka.[5] Again we see that coal prices are far less volatile than those for petroleum products. It may also be seen that the Sri Lanka burner-tip heavy fuel oil price sometimes falls below the corresponding price of coal, and that the product prices broadly follow the highly volatile marker crude price.

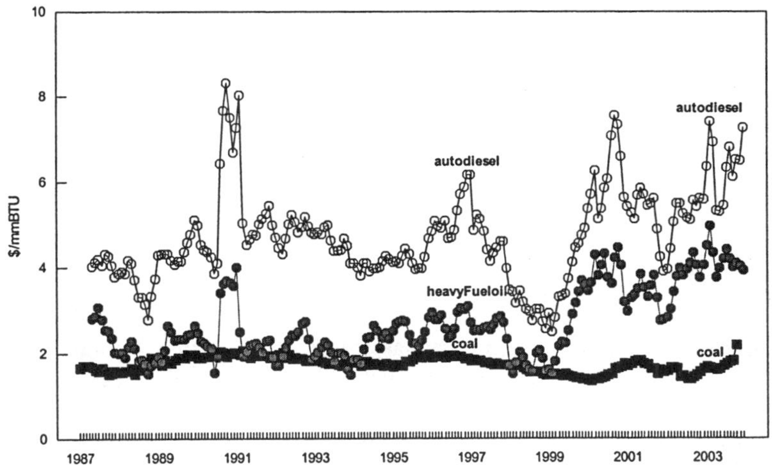

Source: Platts Oilgram and Petroleum Intelligence Weekly, adjusted for freight to Sri Lanka

Figure 4.3 Power generation fuels, cif Sri Lanka

Box 4.1 Coal transportation costs

A study by Connell-Wagner (2001) on coal freight costs to ports in Southern India provides interesting findings. The locations studied were Mangalore, Karnataka, on the west coast, the site for a controversial 1,000 MW coal-fired IPP project (sponsored initially by Cogentrix of the USA, and often referred to as the 'Cogentrix project'), and Madras (Chennai). As shown in Table A, differences between vessel size (US$1.30/tonne between Capesize and Panamax to Madras) is greater than the incremental difference in distance (US$0.70/tonne between Madras and Mangalore for the extra 500 nautical miles). The cost of Australian coal to Sri Lanka from Australia would be comparable to those to Madras.

In fact, the economic cost of imported coal to Mangalore is lower than that of Indian coal (that has to be shipped long distances from fields like Talcher, in Orissa, and that also require rail transport from mine to Paradeep, then a long coastal journey around Sri Lanka to reach India's west coast. The Government of India, reluctant to expose Coal India to international competition, has therefore imposed very high import duties (currently 29 per cent), which effectively preclude use of imported coal except for blending as may be required for environmental reasons.

Table A Ocean freight costs to India, $/ton

Discharge Port	Indonesia Tanjung Bara	Australia Newcastle NSW	S.Africa Richards Bay	China Qinhuang dao
East coast				
Km to Madras (Chennai)	4,300	8,800	6,600	6,900
Handymax	9.50	16.50	13.25	14.00
Panamax	7.50	12.50	10.00	10.50
Capesize	7.25	11.00	8.50	
West coast				
Km to Mangalore	5,100	9,100	5,900	7,700
Handymax	10.50	17.00	11.90	15.50
Panamax	8.50	13.00	8.80	11.00
Capesize	8.00	11.70	7.40	

The relationships between oil and coal price have not been sufficiently appreciated in most analyses of power system options in Sri Lanka. The presumption appears to be that since coal prices are less volatile than those of oil, they are also more or less constant.[6]

The annual generation expansion planning studies of the Ceylon Electricity Board illustrate the problems of specifying fuel price scenarios.[7] Table 4.1 summarizes the assumptions made for fuel prices in the 1992, 1995 and 1998 studies. These represent the input assumptions for the modelling studies, which used the WASP-III capacity expansion planning model.

Table 4.1 Fuel price assumptions in CEB generation planning studies (cif West Coast)

	Base case (constant prices in all cases)	Alternative fuel price scenario
1992 study (1993–2007)	Auto diesel: US$ 16/bbl Coal: US $47/tonne	Auto diesel: US$ 24/bbl Coal: US$ 50.6/tonne
1995 study (1996–2010)	Auto diesel US$ 14/bbl Coal: US$ 48/tonne	Auto diesel increases to US$16.7/ bbl by 1999, then constant. Coal increases to US$ 50.6/tonne by 2000, then constant(a)
1998 study (1999–2013)	Auto diesel: US$ 24.2/bbl Coal: US$ 47/tonne	Auto diesel: US$ 19.7/bbl Coal: US$ 47/tonne

Note: (a) This set of assumptions is described as the 'ADB fuel price forecast', though a specific source for this is not given.

Source: CEB, Report on Long Term Generation Expansion Planning Studies, Generation Planning Branch, Colombo (Reports dated October 1992, July 1996 and January 1999)

Even as deterministic specifications, the fuel price scenarios leave much to be desired. In Figure 4.4, the assumptions of these planning studies are superimposed on the historical relationship between auto-diesel and coal prices, cif Sri Lanka West Coast.[8]

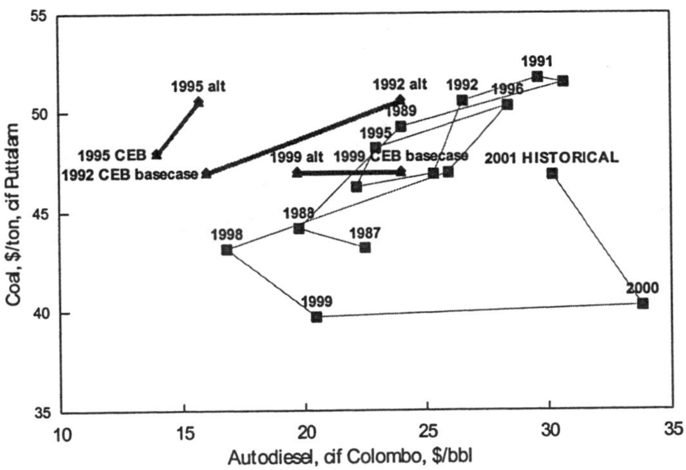

Figure 4.4 Historical relationship between coal and autodiesel prices, cif West Coast Sri Lanka (as annual averages)

Clearly, the 1992 and 1995 studies were very far from any historical relationship of the relative prices of these two fuels. The 1999 base case is consistent with the long term trend-line of coal and auto-diesel prices, but the 'alternative' fuel case, with a lower oil cost but no change in coal price, is again inconsistent.

4.2 LNG

Historically, LNG prices in the Asia-Pacific region have been benchmarked against those for Japan, the largest LNG importer, and which in turn have been tied to crude oil prices (of the 'Japan Crude Cocktail' - JCC) but smoothing out the short-term volatility. Figure 4.5 shows annual average prices, which suggests that the average annual LNG price for long-term planning studies can reasonably be projected as tied directly to the average annual crude oil marker price.[9]

However, the traditional pricing arrangements are undergoing change. LNG already commands lower prices in the more competitive European market, and the traditional long-term contract arrangements for LNG shipments to Japan and Korea based on the JCC are being gradually supplemented by greater reliance on spot and short-term arrangements.

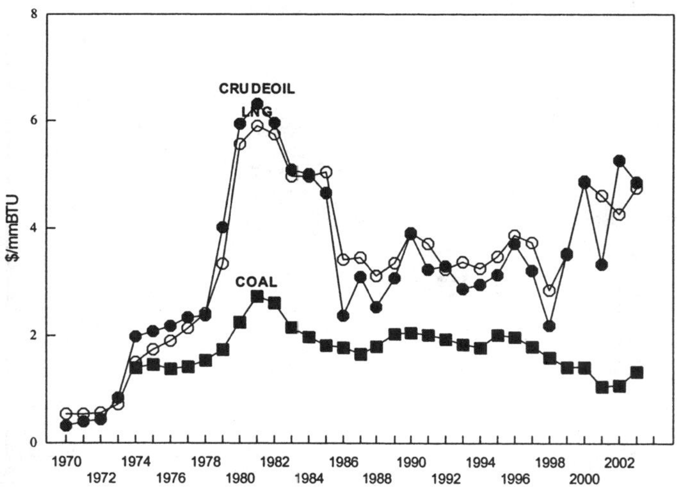

Figure 4.5 Japanese LNG prices v. crude oil and coal

4.3 PRICE VOLATILITY

In addition to concern about the level of prices, over the past decade there is abundant evidence that fossil fuel prices have become more volatile. How to hedge against this volatility, and the costs of doing so, is receiving much attention. Countries such as Sri Lanka, with no domestic fossil resources, are particularly exposed, because among internationally traded petroleum products, auto-diesel is one of the most volatile. Heavy fuel oil and coal are much less volatile, and thus one of the consequences of the failure to implement the coal project is an increasing exposure to oil price volatility as the power sector has become increasingly reliant on auto-diesel as a power generation fuel.

Figure 4.6 illustrates the extent of the problem: the figure shows the monthly and annual average costs of electricity generation in a (notional) 150 MW auto-diesel fuelled combined cycle plant, using actual historical data from 1988 to 2001. Sri Lanka now has 750 MW of oil-fired plants in operation (or under construction) and is heavily exposed to fossil fuel price volatility.[10] Gasoil is assumed purchased on the Singapore spot market. Monthly production costs ranged from a high of almost 8 US cents/kWh during the Gulf Crisis, to a low of 3.4 US cents/kWh in late 1998 when Brent was trading at its recent low of US$ 10/bbl. Even on an annual average basis, costs have ranged from 3.8 cents/kWh to 6.2 cents/kWh.

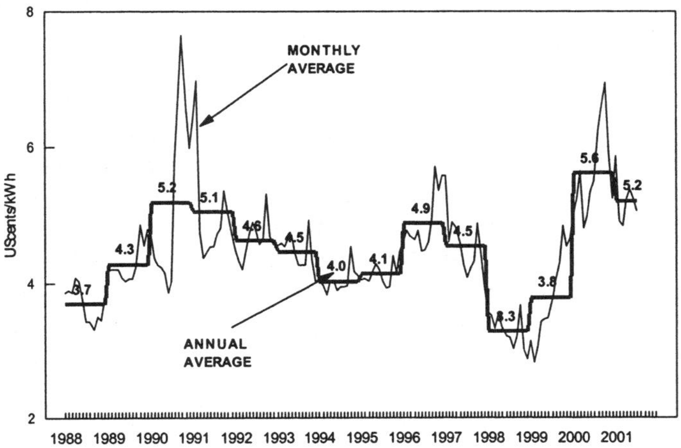

Figure 4.6 Monthly electricity cost of 150 MW auto-diesel CCCT

The consequences of such volatility are rarely examined (and often not even acknowledged) in the relevant cost studies: the topic was not mentioned in any of the generation planning studies of the Ceylon Electricity Board in the 1990s, nor in the IFC study of the proposed 150 MW IPP sponsored by the American company AES (IFC, 1998). For AES as the developer and main equity investor, the issue is relatively unimportant because the contemplated power purchase agreement (PPA) simply passes the fuel price risk to the buyer (CEB). However, for the buyer, such volatility may bring significant financial problems, particularly in the absence of a stable regulatory regime for tariff adjustment.

Renewables as a Hedge Against Fossil Price Volatility

Over the past few years, many advocates have argued that renewable energy serves as a hedge against fossil price volatility (Papathanasiou and Anderson, 2001; Sustainable Energy Advantage, 2003), and that, rather like the role of Treasury bonds in a financial portfolio, an optimal generation portfolio should contain some proportion of renewables on grounds that its costs are less volatile (Awerbuch, 2000; Awerbuch and Berger, 2002).

Because the capital cost is known in advance, but the costs of fossil fuels are not, renewables' generation costs are often claimed to be more predictable and stable (Awerbuch, 2000). However, the relevant question is whether the cost of hedging through renewable energy is also lower than that of using alternative approaches (for example, by hedging with futures or options),[11] or even hedging with other fossil fuels (such as coal) whose price volatility is significantly less than that of oil.

Unfortunately, renewable generation has its own set of unique uncertainties, which cannot always be diversified. For example, some preliminary analysis of wind-farm locations in China suggests strong correlation of monthly output even across wide geographical locations, so building many wind farms in different locations does not diversify the relevant risk (Meier, 2004). Moreover, the value of the hedge provided by renewables compared with those available on the markets has not been tested. While some of the natural variability in wind output can easily be absorbed in systems with significant hydro storage (as in Sri Lanka), the variability in annual production can be significant: studies suggest that the annual variability of generation may be on the order of 10–15 per cent year on year (Garrad Hassan, 1999; Durstewitz et al., 2001).[12]

Figure 4.7 shows the annual average wind speeds reported in the feasibility study for a 3 MW pilot project in southern Sri Lanka, whose long-term average was estimated at 5.74 m/s. The winning bidder for the turbine contract, the manufacturer MICON, estimated the long-term average at the actual site, and at the 46 metre hub height suitable for its 600 kW machine, at 6 m/s. But the actual average wind speed over the first four years of production was only 5.29 m/s, which resulted in an average capacity factor of just 14.2 per cent, as opposed to the 18.7 per cent estimated at time of contract award (Meier, 2003c). Power from this project to date is therefore 35 per cent more expensive than anticipated.

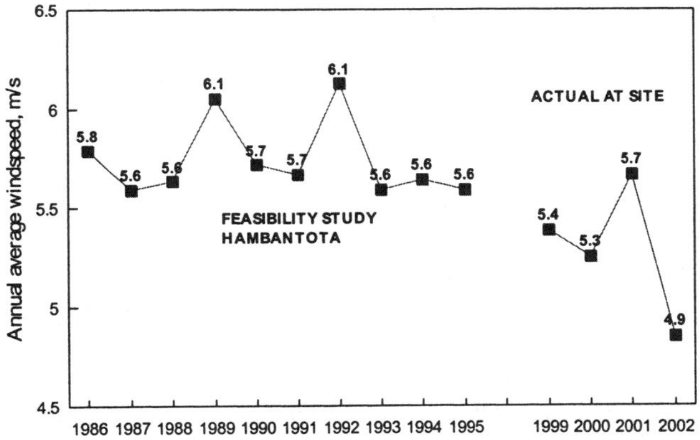

Figure 4.7 Annual average wind speeds: Sri Lanka 3 MW demonstration project

So how does the price volatility of a wind project compare to that of an unhedged autodiesel-fuelled CCCT? This can be investigated by hypothesizing a wind plant of the same average annual energy production as the 150 MW CCCT plant shown in Figure 4.6 (though ignoring, for the moment, the problems of matching the seasonal and daily load curves by a less-dispatchable technology).[13] In this example we assume a wind plant with a long-term average 31 per cent capacity factor, and a capital cost of US$750/kW, chosen so that the expected levelized economic cost of both wind and CCCT is 4.4 US cents/kWh. Both are assumed to have the same annual electricity output of 1,215 GWh/year. We assume further a full capacity credit for wind. These would certainly be 'best case' assumptions for wind power in a comparison with oil-based plant.

Figure 4.8 shows the result of a typical simulation for the case that the annual variation of wind output has a coefficient of variation of 10 per cent, as suggested by Garrad Hassan (1999): one would be hard-pressed to argue that the variations in wind generation costs were smaller than those of the corresponding fossil plant.

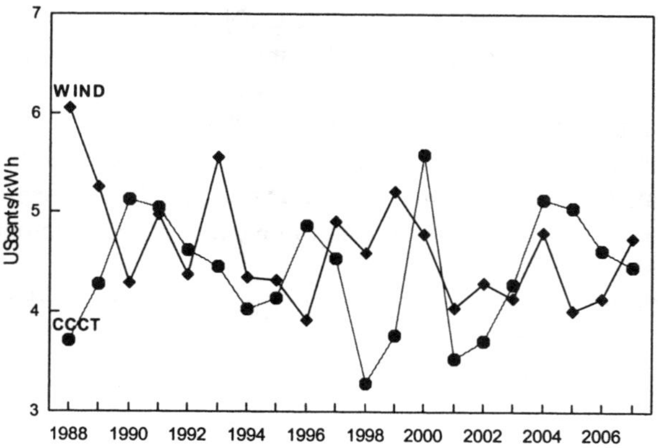

Figure 4.8 Comparison of annual variations in costs

There is the additional problem of capacity penalties that apply to non-dispatchable renewable energy generation – penalties that arise because wind energy may not be available during peak hours, which therefore requires additional standby capacity to meet peak loads. The extent of penalties has been vigorously debated in a growing literature. Estimates of the capacity credit in a range of US systems have been reviewed by Grubb and Meyer (1992). The capacity credit was found to generally decrease as the level of wind in the system increases. For example, in the Kansas gas and electric system the capacity credit falls from 50 per cent at 5 per cent wind penetration (wind MW as a percentage of system peak) to 30 per cent at 20 per cent penetration. At low penetration levels (5–10 per cent), most estimates of capacity credit are between 20–50 per cent, though one study of the PG&E system suggested a capacity credit of 70 per cent.

More recent analyses (Milligan and Parsons, 1997; Milligan, 2003) note the difference between capacity penalties for planning and for operations. The former looks at the problem from the perspective of investment planning: the capacity value of a wind plant must be quantified so that investment of conventional generating capacity can be partially offset by the capacity value of the wind plant. The latter looks at the problem from the perspective of the dispatcher, who must schedule conventional resources to ensure adequate generation to meet the load on an instantaneous basis.

In general, the penalty increases as the proportion of wind energy rises; and decreases with the amount of hydro-storage in the system (particularly if the hydro system is energy rather than capacity constrained). In a detailed study of the wind capacity penalties in the North China grid (see Box 4.2), 2,600 MW of wind power displaced only some 1,130 MW of fossil-based energy.

Thus the extent to which renewable energy can serve as a hedge against fossil fuel price volatility remains to be clearly established, and is the subject of an ongoing World Bank research project. Certainly in the case of wind power in Sri Lanka, sites must be found with much better average annual capacity factors than that of the demonstration project before one can talk of a meaningful hedge against the uncertainties of fossil price variations.

Box 4.2 Wind capacity penalties in the North China Grid

The economic analysis of the China Renewable Energy Scaleup Program (CRESP) examined the impact of a mandated market share approach to increasing renewable energy generation in China (Meier, 2004). The best way of establishing the extent of capacity penalties is to run a power systems' generation capacity expansion planning model, comparing a free run with that into which given quantities of wind energy are forced into the system.

Using the GESP model developed by the Beijing Power Economic Research Centre, some 2,600 MW of wind power (located mainly in Inner Mongolia) was forced into the North China Grid (which serves a series of major metropolitan areas including Beijing) in a maximum wind power scenario. The result was that this displaces only 1,132 MW of coal and oil-fired CCCT capacity, for an overall capacity credit of around 43 per cent. A similar study was conducted for a small hydro scenario in Zhejiang province, for which the small hydro was found to have a slightly higher capacity credit of 46 per cent.

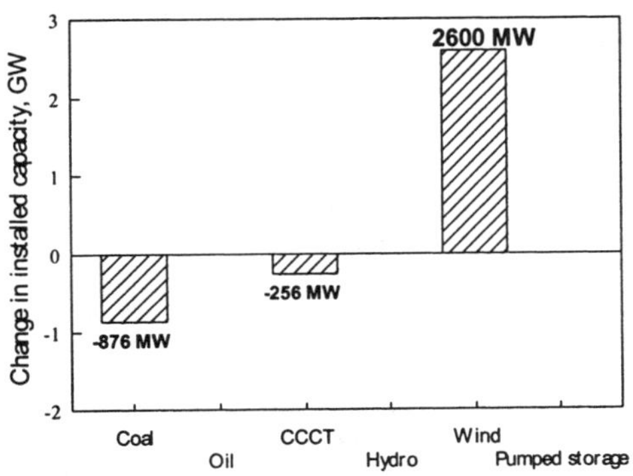

NOTES

1 Proposals for coal-fired IPP projects in South India, such as the 2 × 500 MW project at Mangalore in coastal Karnataka, now generally propose the use of a mix of South African, Indonesian and Australian coals in their feasibility studies.

2 The trend line used for modelling purposes (bold line in the figure) is estimated as

$$Coal[fobNSW,\$/tonne] = 21.9 + 0.73[Brent,\$/bbl];\ \mathrm{R}^2 = 0.854$$

This excludes the years 1984-85, and 1999-2001, for reasons discussed in the text. If these years are included, the relationship changes to *coal=25.29+0.47[Brent]*, with a significantly reduced R^2 of 0.33.

3 The term 'marker' crude derives from its use in pricing formulae for other types of crudeoils. For example, Kuwait produces an export blend (31 API, 2.5 per cent), for which the formal marker is the monthly average of the Dubai + Oman averages, plus a differential that in 1996 ranged from US$-0.3/bbl to US$-0.7/bbl.

4 There are many reasons why Brent crude is increasingly appropriate as a marker for the Asia-Pacific market. First, over recent years there have been significant flows from Nigeria, Angola and Gabon, at prices linked to Brent, into Asian markets (largely in response to the need for lower sulphur fuels, given lack of desulphurisation capacity in many Asian countries, particularly Korea). Second, VLCCs serve as an effective arbitrage mechanism against attempts of Middle Eastern exporters to discriminate between eastern and western flows. Third, liquidity of the Dubai market is thin, with relatively few trades reported. Finally, both Dubai and Brent are predominantly traded in London, with largely similar participants. For further details, see Horsnell (1997).

5 Coal transportation costs to Trincomalee would be less because larger vessels could be used, making a difference of about US$ 3/tonne.

6 For example, in the sensitivity analysis reported in the 1998 CEB expansion plan (CEB 1998b), an expansion plan was derived for a low oil price case, with a 20 per cent reduction in oil prices from US$ 18.5/bbl for Saudi Light, but keeping the coal price constant. However, in 1997 the corresponding sensitivity analysis was for a 10 per cent increase in crude price accompanied by a 5 per cent increase in coal price. This is seen to be a more realistic case, given the relationship noted above.

7 It should be noted that the severe power shortages experienced in Sri Lanka in 1996 and 2001/02 were the consequence of decision-making paralysis in the Government, not shortcomings of CEB's planning studies – which, if they had been implemented, would have added the necessary generating capacity and avoided the power cuts. Claims that the power cuts were the consequence of unexpected 'drought' were simply incorrect: in fact the explanation was the failure to implement the coal-fired plant (recommended for implementation as far back as 1990), instead relying on *ad hoc* additions of combustion turbines when crisis loomed.

8 Auto-diesel assumed at Singapore fob plus US$ 10/tonne freight. Coal as fob Newcastle(Australia) plus US$ 12.50/tonne freight (based on a detailed study of coal freight costs to southern India by Connell Wagner, 2001)

9 For details of LNG price formation, see for example, Fujime (2002): the so-called 'Asia' premium of about US$ 1/bbl (for both oil and LNG) is attributed to long-term contracts linked to average Dubai and Oman prices. In addition, higher LNG prices have been paid because of the linkage to crude oil, rather than burner-tip prices with immediate substitutes. Thus LNG prices have been as much as US$ 1/mmBTU higher than those paid by US and European LNG consumers. The average 2000 CIF price of LNG imported into the USA was US$ 3.18/mmBTU, compared to US$ 4.88/mmBTU in Japan.

10 Including a 60 MW barge-mounted plant in Colombo Harbour, burning fuel oil; 300 MW of CCCT, 44 MW of old steam cycle plant, and 235 MW of simple cycle gas turbines (all using auto-diesel and sited at Kelanitissa); and a variety of small IPP diesel units in smaller towns totalling 110 MW. Another 300 MW of auto-diesel fuelled CCCT is planned for Kerawalapitiya (north of Colombo).

11 For an example of oil futures hedging in a small, oil-importing developing country (Costa Rica), see, for example, Myers and Thompson (1993).

12 When expressed as the coefficient of variation, that is, standard deviation divided by the mean.

13 Wind and run-of-river small hydro are often referred to as non-dispatchable. However, thermal plants also have forced (and maintenance) outages. Indeed, while thermal forced outages may be assumed to be completely random, in the case, say, of wind power, there are clear daily and seasonal patterns to the variation in output (though wind plants are also subject to forced outage attributable to equipment failures).

5. Resolving technology and fuel choice problems for power generation

For more than a decade Sri Lanka has struggled with fundamental choices necessary to secure the nation's future electricity supply. In 1998, the Government and the Ceylon Electricity Board (CEB) requested assistance from the World Bank in providing an economic and financial analysis of technology and fuel options. Its objective was to review the technology and fuel choices available to Sri Lanka, and to document the economic and financial aspects of these options to a similar level of detail, in order that decision-makers could judge for themselves the advantages and disadvantages of each. This chapter summarizes the results of that study (Meier and Munasinghe, 2003c).

5.1 ASSUMPTIONS AND APPROACH

The conventional practice in economic analysis is to price projects funded in the public sector at the real social discount rate (SDR), which is taken here at 10 per cent. This is also the rate used by CEB in its generation planning, as well as by CEB's consultants for benefit–cost analysis in detailed feasibility studies. This rate is at the lower end of discount rates currently used in Asia: the rate in the Philippines is 15 per cent, China 12 per cent and Vietnam 10 per cent.[1]

However, in the case of privately financed projects, particularly foreign-financed IPPs, the actual cost of capital may be significantly higher. For example, the weighted cost of capital for the 250 MW oil-fired project that had been proposed by the Electricity Corporation of New Zealand (ECNZ) in the late 1990s was over 20 per cent. However, in others (such as the AES combined cycle project), the apparent weighted cost of capital is fairly close to 10 per cent. The following technology-fuel options were evaluated:

- Steam cycle plants: coal and imported heavy fuel oil
- Diesels: residual oil, imported heavy fuel oil, diesel oil
- Combustion turbines: diesel oil, naphtha, LNG
- Hydro conventional and mini-hydro
- Renewables: wind, dendro-thermal

A preliminary screening identified a number of technology fuel options that are not likely to be viable in Sri Lanka, and therefore not examined in great detail. A first group represents technologies that have been advocated for Sri Lanka, but which may be eliminated from consideration within the next decade, since they are quite far from the stage of being commercially proven or are economically unviable – this group includes wave and tidal energy, ocean thermal energy conversion (OTEC) and solar thermal, and most of the so-called 'clean coal' technologies, including in particular integrated gasification-combined cycle (IGCC) and pressurised fluidised bed combined cycle (PFBC).

This is not to say that they have no long-term potential, or that they are not appropriate as subjects of research. However, it does say that these options are not candidates for the short to medium term, and technological progress in any one of these technologies does not affect the decisions that must be taken over the next few years. In the case of IGCC and PFBC, neither can yet be viewed as commercially available. PFBC is presently offered by only a single vendor, and has been applied elsewhere only in developed countries in situations of extraordinarily stringent ambient air quality requirements.

A second category of technologies may be deemed to be technologically proven and commercially available, but their advantage is limited to circumstances that simply do not apply to Sri Lanka. These include:

- Nuclear: on grounds of cost and unit size.
- Atmospheric fluidised bed combustion (AFBC): of interest primarily where a country has large domestic resources of poor quality solid fuel (such as India and China).
- Liquid petroleum gas (LPG): for large plants on grounds of high fuel cost, for small plants on grounds of cost and limitations in import infrastructure.

However, since clean-coal technologies have been examined in some past studies of Sri Lanka, a summary analysis of AFBC is provided for comparison of costs and emissions of other baseload options.

Two options were analysed in particular detail, namely oil-fired steam cycle plants (an interest sparked by the ECNZ proposal) and LNG (for which ENRON had made a proposal to CEB in 1999). The ECNZ project involved the dismantling of the 250 MW oil-fired Marsden B station and its re-installation in Sri Lanka: this plant was mothballed in New Zealand in 1979 just prior to commissioning following the discovery of large quantities of natural gas. Although this project has now been rejected, it provides a useful case study of how difficult it is for governments to evaluate IPP proposals that are not competitively bid.

We then present a financial analysis, in which the impact of assumptions about the financing package are evaluated. These prove to have a far greater impact on the results than the distortions imposed by fuel taxes and excise duties on equipment.

Needless to say, the interest in oil-fired projects occurred at a time when world oil prices were extremely low (having reached a historic low of US$10/bbl in late 1998). In 1999, as oil prices began to increase, the prevailing conventional wisdom was that world crude oil prices might well reach the earlier trading range of US$15–20/bbl, so an Arab Light price of US$18–20/bbl was considered 'high' (particularly by the advocates of oil-based generation projects).

From the perspective of US$35–50/bbl oil prices experienced in 2004, a discussion of oil-based electricity generation may seem academic, particularly after several years of historically low coal prices (relative to oil, as noted in the previous chapter). But between mid-2003 and mid-2004, coal prices in the Asia-Pacific market have almost doubled, so again the cost differences between coal and oil-fired generation have narrowed. Nevertheless, as shown in the concluding section of this chapter, where we update the analysis to mid-2004 conditions, the recommendations remain unchanged.

5.2 SCREENING CURVE ANALYSIS

Simple one-on-one comparisons of different technologies were undertaken using a screening curve model that assessed cost/kWh and lifetime net present value (NPV) as a function of annual plant factor. Although most assumptions were based on the 1998 CEB generation plan and project feasibility studies prepared by CEB's consultants, these have been modified as necessary in the case of inconsistencies and other anomalies identified.[2]

Table 5.1 lists the assumptions for the one-on-one comparisons based on an US$18.5/bbl Saudi Light price (about US$20.4/bbl for Brent). It is assumed that oil-steam uses imported fuel oil blended to 1.6 per cent sulphur.[3] The indicated capital costs are adjusted for opportunity cost of capital and the infrastructure penalty.[4]

Table 5.1 Input assumptions

		Pulv. coal	OCCT	CCCT	Diesel (resid)	Diesel (barge)	Oil-steam
Total capacity	MW	300	105	300	150	300	300
Capital cost	US$/kW	1,303	386	725	1,332	1,334	1,063
Life	Years	30	20	30	25	25	25
Fixed O&M	US$/kW/month	0.56	0.36	0.28	0.92	0.92	0.48
Variable O&M	mills/kWh	2.79	2.85	2.72	7.36	7.36	2.37
Sched. maintenance	Days	40	30	30	30	30	30
Forced outage rate		0.03	0.08	0.08	0.15	0.15	0.06
Heat rate, net	KCal/kWh	2,293	3,060	1,890	1,954	1,954	2,293
Primary fuel		coal	auto diesel	auto diesel	resid	furnace oil	Imp. fuel oil
Heat content	KCal/kg	6,300	10,550	10,550	10,300	10,300	10,300
Fuel price, cif	US$/bbl	0	28.1	28.1	11.6	15.3	15.8
	US$/ton	49	212	212	78	103	106
	US$/mmBTU	1.96	5.05	5.05	1.91	2.51	2.6
	cents/Gcal	779	2,006		756	997	1,031
Fixed charge factor		0.11	0.12	0.11	0.11	0.11	0.11
Annual capital cost	US$/kW/year	138.2	45.3	76.9	146.7	147	117.1
Fixed O&M/year	US$/kW/year	6.7	4.3	3.4	11	11	5.7
Total fixed cost	US$/kW/year	145	49.6	80.2	157.8	158	122.9
Fuel cost	US cents/kWh	1.79	6.14	3.79	1.48	1.95	2.36
Variable O&M	US cents/kWh	0.28	0.29	0.27	0.74	0.74	0.24

The screening analysis showed coal to be the least cost for baseload duty, combined cycle for cycling duty (PLF 15–40 per cent), and open cycle combustion turbines for thermal peaking (Figure 5.1). However, at an US$ 18/bbl oil price, the advantage of coal over oil-steam is small, and in the range of 40–50 per cent PLF, essentially identical. When the analysis is repeated for the lower oil price of US$ 16/bbl, the differences between coal and oil-steam are smaller still.

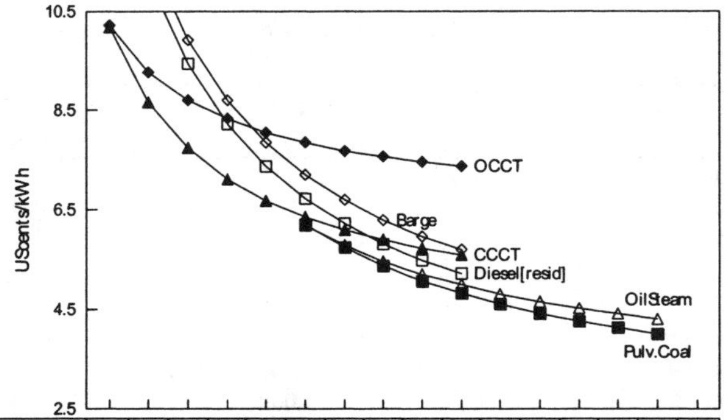

Annual PLF		0.20		0.30		0.40		0.50		0.60		0.70		0.80		0.90
■ Pulv.Coal						6.2	5.7	5.4	5.1	4.8	4.6	4.4	4.3	4.1	4.0	
◆ OCCT	10.2	9.3	8.7	8.3	8.0	7.8	7.7	7.6	7.5	7.4						
▲ CCCT	10.2	8.6	7.7	7.1	6.7	6.4	6.1	5.9	5.7	5.6						
⊟ Diesel	14.2	11.2	9.4	8.2	7.4	6.7	6.2	5.8	5.5	5.2						
◇ Barge	14.7	11.7	9.9	8.7	7.8	7.2	6.7	6.3	6.0	5.7						
△ OilSteam						6.2	5.8	5.5	5.2	5.0	4.8	4.7	4.5	4.4	4.3	

Figure 5.1 Screening curve (Arab Light 34 US$18.5/bbl)

The differences between technologies are further illuminated by looking at the lifetime cost differences between coal and the other alternatives for some constant plant load factor – as shown in Table 5.2. The advantage of coal over steam-cycle oil is in the range of US$ 8–37 million (and at today's US$ 35–45/bbl oil price, over US$ 65 million).

Table 5.2 Lifetime cost difference US$ million (NPV at 10 per cent over 30 years) for baseload duty at 75 per cent PLF

	US$16/bbl		US$18.5/bbl	
	NPV	δ (coal)	NPV	δ (coal)
Coal	776		794	
CCCT	886	110	982	188
Diesel (residual)	821	45	857	64
Diesel (furnace oil)	909	133	946	152
Oil-steam	784	8	831	37

Capital cost assumptions are also problematic. In cases where recent Sri Lankan prices have been revealed in international competitive bidding (such as for open cycle combustion turbines), they are relatively firm; in others, such as for oil steam cycle plants (or LNG, discussed below), the estimates are subject to high uncertainty. In such cases, a switching value analysis can provide further insight.

If, for example, the switching value of an alternative to coal requires a more than 86 per cent decrease in capital cost from the baseline to be competitive – as in the case of auto-diesel fired combined cycle for baseload duty (Table 5.3) – one may be reasonably sure that coal is least cost (other things being equal); but if the alternative requires only an 11 per cent decrease (as in the case of oil-cycle steam), then the assumptions for that alternative require further scrutiny.

Table 5.3 Capital cost switching values

	Baseline estimate (US$/kW)	Switching value (US$/kW)	Reduction in baseline to attain the switching value (%)
Coal	1,303		
CCCT	725	98	86.5
Diesel (residual)	1,332	1,127	15.4
Diesel (furnace oil)	1,334	846	36.6
Oil-steam	1,063	944	11.3

Small Diesels

Over the past few years a number of small diesel IPP plants running on diesel oil have been built in Sri Lanka (such as the 22.5 MW Lakdanavi and 50.8 MW AsiaPower projects). Unlike diesels using residual fuels or heavy fuel oil (such as those at Sapugaskanda), such units have significantly lower capital costs – but correspondingly higher fuel costs.

Figure 5.2 shows the screening curve for small diesels (compared to options suited to intermediate and baseload), in which we assume a capital cost of US$600/kW. It is evident that at such capital costs, these plants are uneconomic for baseload operation, but are similar to CCCTs for load factors of less than 40 per cent. Unfortunately, during the power shortages of recent years, these diesels have been run at high load factors, in turn causing severe financial distress at CEB (in 2002, CEB lost Rs 7 billion).

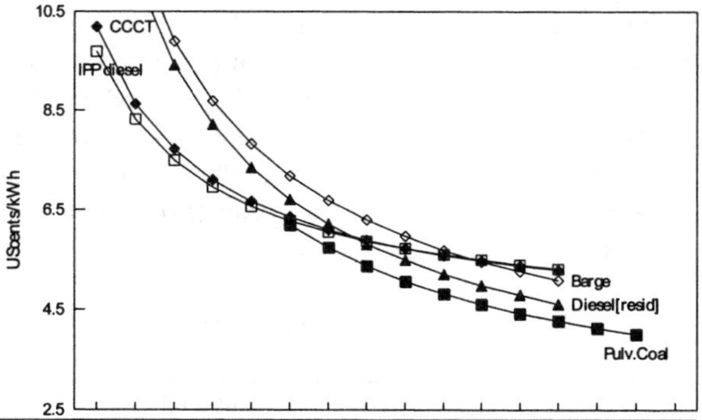

Annual PLF		0.2		0.3		0.4		0.5		0.6		0.7		0.8		0.9
■ Pulv.Coal						6.2	5.7	5.4	5.1	4.8	4.6	4.4	4.3	4.1	4.0	
◆ CCCT	10.2	8.6	7.7	7.1	6.7	6.4	6.1	5.9	5.7	5.6	5.5	5.4	5.3			
▲ Diesel[resid]	14.2	11.2	9.4	8.2	7.4	6.7	6.2	5.8	5.5	5.2	5.0	4.8	4.6			
⊟ IPP diesel	9.7	8.3	7.5	7.0	6.6	6.3	6.0	5.9	5.7	5.6	5.5	5.4	5.3			
◇ Barge	14.7	11.7	9.9	8.7	7.8	7.2	6.7	6.3	6.0	5.7	5.5	5.3	5.1			

Figure 5.2 Screening curves for small diesels
(10 per cent discount rate, US$ 18.5/bbl Arab Light price)

The above analysis uses a 10 per cent discount rate to reflect the opportunity cost of capital. Since generation technologies vary in capital intensity, the results may be sensitive to the rate assumed. In Figure 5.3 we show the lifetime costs for a range of discount rates from 8–14 per cent. Technology rankings do not change until one moves from 12–14 per cent. At 14 per cent, the advantage of coal is very small indeed.

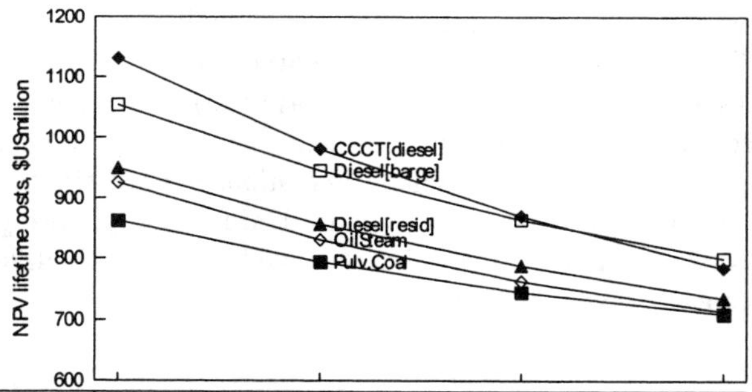

discount rate	0.08	0.10	0.12	0.14
■ Pulv.Coal	862	794	745	710
◆ CCCT[diesel]	1130	982	871	785
▲ Diesel[resid]	950	857	788	736
⊟ Diesel[barge]	1055	946	864	801
◇ OilSteam	925	831	763	713

Figure 5.3 Sensitivity to discount rate: lifetime NPVs
(75 per cent PLF, US$ 18.5/bbl Arab Light), in US$ million

The Impact of Delay

In terms of lifetime NPV costs, the differences between some technologies (for example, between coal and oil steam cycle plants for equivalent baseload duty, a range of US$ 8–37 million) are seen to be relatively small. Equally important, these differences are also relatively small compared to the cost of delay.

Suppose, for example, a one-year delay in making a commitment to a 300 MW base load plant, and that the power lost (1,971 GWh assuming 75 per cent PLF) must be generated in thermal peaking units. Suppose further that the alternative generation capacity has a shorter gestation time and could be advanced from its least cost commissioning date by one year. Then the impact of delay is the difference in fuel cost between generating this 1,971 GWh in the coal plant, versus having to generate this power with diesel oil in either open cycle combustion turbine (OCCT) or combined cycle combustion turbine (CCCT) plants, less the saving attributable to deferring the capital costs of the coal plant by one year, plus the cost of advancing the alternative by one year.[5] As indicated in Table 5.4, this cost is in the range of US$ 24–60 million (in NPV terms).

Table 5.4 Impact of delay

		Coal	OCCT	CCCT
Capital cost	(US$ million)	411	116	217
Capacity	(MW)	300	300	300
PLF		0.75	0.75	0.75
Replacement generation	(GWh/year)	1,971	1,971	1,971
Fuel cost	(US cents/kWh)	1.79	6.14	3.79
Annual fuel cost	(US$ million)	35.2	121	74.7
Delta (coal)	(US$ million)		85.8	39.5
Capital cost impacts				
Delay coal by 1 year	(US$ million)		-35.5	-35.5
Advance alternative by 1 year	(US$ million)		10.5	19.8
Total impact	(US$ million)		60.7	23.7
Advance alternative by 3 years	(US$ million)		28.8	54.1
Total impact	(US$ million)		79.0	58.0

However, this assumes that the adjustment required is to advance the alternative by one year. If that alternative (with shorter gestation time) must be advanced by more than one year (relative to the least-cost expansion plan), then the impact is correspondingly greater: if the advance time is three years, the total impact of delay rises to US$ 58–79 million, significantly higher than the difference between lifetime NPVs of coal and oil-steam (US$ 8–37 million).

Nevertheless, this is not true of all technology comparisons. The lifetime cost difference between an auto-diesel fuelled CCCT and a steam-cycle plant lies in the range of US$ 110–188 million – which is above the cost of a one to two-year delay. In the present highly competitive international generation equipment market, capital costs for all types of power generation plant have fallen dramatically over the past few years (though particularly for combustion turbines, driven in part by technological advances in the related jet engine market). Nevertheless, the switching values analysis for capital costs shows that for baseload duty, only oil-steam running on heavy fuel oil, or diesel using residual oil, are in the competitive range with coal. Since additional quantities of residual oil from the Sapugaskanda refinery would only be available if the refinery were expanded (a highly unlikely event in the short to medium term), large-scale use of such diesels is not feasible for the time being.

It has been the practice of CEB to compare the coal plant against an auto-diesel CCCT to justify the former in economic analysis: the CCCT is claimed to be the 'next best' option. Our results show that the second-best option for baseload is steam-cycle oil (which is true even at higher world oil prices).

It cannot be ruled out that in a competitive bid in which the choice of fuel/technology were left to the bidder, coal would necessarily win. Indeed, there is scope for optimisation and hedging techniques that would be within the competence and experience of an international IPP, which may make oil more attractive to a private owner. But the recent rise in oil prices, makes it increasingly unlikely that IPPs would take on the oil price risk.

If the fuel price risk is taken by the buyer (that is, by CEB or its successor entity), then comparison of bids across different fuels and technologies would involve a risk evaluation, in which the higher fuel price risk associated with oil needs to be traded-off quantitatively (and transparently) against the lower capital cost. Such a risk assessment (and related questions of benefit sharing, structuring and quantifying the corresponding criteria for evaluation of proposals, etc.) falls outside the scope of this book, but should certainly be undertaken before one could consider a request for proposals (RFP) that leaves technology and fuel choice to the IPP.

Finally, the concept of multi-fuel capable plants also merits some further examination – as noted in Chapter 9, co-firing with wood (at a low volume ratio) in the coal project is a potentially attractive GHG mitigation measure.

The Impact of Infrastructure Front Loading

One of the contentious issues in the public debate over new generation plants is the practice of averaging up-front infrastructure costs over all the units of a multi-unit plant in economic analyses. This has been done both by CEB in its generation planning studies, as well as by international consultants in their detailed feasibility studies (such as Tokyo Electric Power for the Kerawalapitiya project, or Electrowatt for the West Coast Coal project). The issue is potentially significant, for the differences between the first and subsequent units is sometimes large. Critics claim that this has biased past studies in favour of coal (which have particularly high up-front costs). For example, as shown in Table 5.5 for the Electrowatt studies, while the first unit cost for coal is US$1,451/kW, the average is US$ 1,008/kW, and the cost for Units 2 and 3 is US$787/kW.

Table 5.5 Capital costs of the coal plant

	Unit 1 at actual	Unit 1 at average	Units 2 & 3 at actual
Outdoor site works	38.1	17.1	6.7
Buildings and structure	32.9	26.3	23.0
Steam generation plant	82.4	80.5	79.5
Turbine generator plant	60.0	58.6	57.9
Fuel and ash handling	25.5	12.2	5.5
Balance of plant	32.7	20.0	13.8
Switchyard equipment:	19.0	15.4	13.7
Control and instrumentation	6.2	5.6	5.3
Resettlement costs, environment, monitoring	1.5	0.5	0.0
Subtotal 1	298.2	236.3	205.3
Contingencies	14.9	11.8	10.3
Engineering and supervision	14.9	11.8	10.3
Subtotal 2	328.0	259.9	225.9
Coal unloading	65.3	21.6	
Transmission	41.9	21.0	10.5
Total	435.3	302.4	236.0
US$/KW	1,451	1,008	787

Source: Detailed feasibility and economic analysis studies by Electrowatt

Use of the average value – in either system planning models or simple one-on-one comparisons – results in an effective under-statement of project capital costs of about 7 per cent in the case of the coal project, with the second and third units commissioned three and seven years after the first unit (IDEA, 2000). Clearly, use of the first unit costs as a basis for decisions would give an equally biased picture, since subsequent units have substantially lower costs.

However, increasing capital costs by 7 per cent has little impact on the results – we find that coal is still least-cost for baseload requirements. Nevertheless, we recommend that in future WASP generation planning studies undertaken by CEB, such an adjustment be made. To be sure, if only the first unit at a multi-unit plant is built, then its capital costs are effectively 40 per cent higher than the assumed average. However, it is very unlikely that Units 2 and 3 would not be required. More specifically, there are only three circumstances under which a coal project might be abandoned after the first unit.

The first possibility is that load growth would fall from the expected 7–8 per cent to, say, only 2–3 per cent, circumstances under which the second and third units might see very long postponements. However, even if there is the occasional year that deviates from the long-term trend (as occurred in 1996), the probability that there will be a marked deceleration of load growth implies a corresponding decline of the long-term GDP growth rate. The latter outcome seems very unlikely given the consensus among all political parties for pursuing economic growth through liberalization and reform, and, notwithstanding differences among the parties in negotiating approach, to the peace process with the Tamil Tigers. Over the longer term, a sharp acceleration of the electricity growth rate may be expected once a sustainable settlement of the ethnic conflict allows full reconstruction of the disturbed areas in the north and east of the country.

A second reason why coal-fired units 2 and 3 would not be built is a fundamental change in the structure of international energy prices, under which oil (and LNG) would become much cheaper than coal. However, as noted, coal prices are not unrelated to oil prices over the long term. Thus, any prolonged decrease in oil prices is inevitably accompanied by a corresponding decline in coal prices, just as a prolonged increase in oil prices will bring an increase in coal prices (though, as shown over the past few years, it may take several years for the coal to oil price equilibrium to be restored).

Even if over the next ten years a global carbon tax regime came into place, it is entirely unclear that this would necessarily result in a decline in oil prices relative to coal. While unit rates of carbon tax would certainly be higher on coal than on oil (and gas), coal producers are just as likely to reduce prices to maintain market share, while oil producers may increase prices to capture the available premium. In any event, the impact of a plausible level of carbon tax is likely to be far smaller than the potential impact of geopolitical problems in the Middle East, concerted action by OPEC, or the full scale resumption of Iraqi oil production.[6]

The third and final reason for concern would be the unexpected discovery of large natural gas resources in Sri Lanka – which would put Sri Lanka in the same situation as New Zealand in 1979 (which, as noted, resulted in abandonment of the Marsden B station). However, if indeed this were to occur in Sri Lanka, the benefits to Sri Lanka of large natural gas fields would be so overwhelming as to make abandonment of the coal plant a relatively minor matter (again, exactly as in New Zealand with Marsden B). Moreover, if units 2 and 3 at Puttalam were to be fuelled by gas, some part of the site infrastructure would still be usable (though not of course the coal unloading and handling facilities).

In short, it is highly unlikely that if indeed a decision to proceed with coal were taken (or if coal emerged as the least cost fuel in a competitive solicitation for baseload power), that the project would be abandoned after the first unit.

5.3 LNG

Most detailed engineering studies undertaken by independent consultants reject LNG as a viable option for power generation for the time being, on both economic and technical grounds. For example, Tokyo Electric Power Services (1998), one of the world's largest and most experienced users of LNG (and therefore hardly likely to have a bias against LNG), concluded in their feasibility study of the proposed west coast 750 MW Kerawalapitiya project that the use of LNG would effectively double the capital cost, and that it was substantially impossible to unload LNG at west coast sites during the monsoon season.[7]

However, LNG has had vocal advocates in Sri Lanka. Indeed, there is no question that from an environmental point of view, LNG is an attractive option when compared to coal or oil-based generation plants. Moreover, LNG has the advantage that it is readily burnt in combustion turbines (with or without a combined cycle) that are characterised by high efficiency and which have seen significant recent decreases in capital costs.[8]

But whether in the case of Sri Lanka these advantages offset the disadvantages, notably the very high cost of LNG, is unclear. In the late 1990s the most vocal advocate of LNG was ENRON, which made a proposal to CEB in June 1999. While the proposal contained much marketing rhetoric,[9] it was distinctly short on the specifics of the actual price of LNG or power being offered. Some of ENRON's views about project gestation periods bordered on the wishful thinking, claiming that an 'LNG/interim fuel' project required only 2-3 years to start up as opposed to 4-5 years for a coal project. Obviously one can build a distillate-fueled CCCT in 2-3 years. But there is nothing in ENRON's Indian experience that suggests an LNG project can be quickly implemented: Phase I of the Dabhol project, which used naphtha and distillate as 'interim' fuels, in fact required four years from construction begin to commissioning, and seven years from signing of the MoU.[10]

Non-power Uses

In order to benefit from the economies of scale of LNG, it is clear that a project of economic size in Sri Lanka would require substantial non-power gas use. However, unlike other countries in South Asia (notably India), which already have a gas pipeline and distribution system, huge investments would be required to establish a new system. Sri Lanka also lacks large energy-intensive industries that could easily switch from oil to gas were the latter to become available.

To be sure, it may be argued that every country that now has a domestic gas pipeline system had to begin at some point. However, we know of no case anywhere in the world where such a pipeline system was started on the basis of LNG imports. In every example known to us, the impetus for a gas pipeline system was the discovery of domestic gas reserves (as in the case of India and Pakistan), or access to gas pipelines fed by gas-rich neighbours (as in the case of Switzerland supplied from North Africa via Italy, or Spain from Algeria via Morocco). LNG is readily injected into an existing pipeline system (as for example in the US, where east coast LNG projects were seen as a way of meeting high-value winter heating demand peaks, or as in many of the proposed Indian LNG projects),[11] but a gas pipeline system based entirely on imported LNG is hard to envisage.

A direct pipeline from the Gulf can be ruled out (Oman to Colombo is about 2,700 km). Even a gas pipeline from a southern Indian LNG terminal implies a distance of some 500-800 km, which for the sorts of volumes likely to be required in any first stage would be very expensive.[12] Moreover, even if gas were discovered in significant quantities off-shore in south India, there is no reason to believe that the fob gas price to Sri Lanka would be at a substantial discount to international price levels. Were gas to be discovered in Sri Lanka waters, the situation is of course arguably different: but there is presently no evidence of commercially exploitable gas reserves in Sri Lanka.

In countries with acute water supply problems, desalinisation represents a potential high-value gas use that could be integrated into a coastal LNG project – as developed by ENRON in Puerto Rico.[13] However, this use is not indicated in Sri Lanka.[14]

It has been suggested that compressed gas might be used in Sri Lanka for transportation use, for indeed this would certainly represent a substantial market. However, even leaving aside for the moment the economics, it may be noted that the total transportation use of diesel and gasoline in Sri Lanka is about 1.2 million tonnes per year (mtpy).[15] With a 300 MW power project using about 0.3 mtpy, even a 2 mtpy LNG project would still require a non-power market for 1.7 mtpy, substantially greater than the total present transportation demand in the entire country. Again we make the observation that we know of no country anywhere in the world where large-scale use of compressed natural gas in transportation has been undertaken in the absence of large scale domestic gas reserves. Both New Zealand and Thailand, which have large numbers of CNG vehicles, have substantial indigenous natural gas reserves: neither imports LNG for this purpose.

Finally there is the possibility of converting other gas turbine power plants from liquid fuels to gas – of which as much as 1,000 MW might be in place by the end of the decade. But the infrastructure costs of converting these plants would be high, since it would require a pipeline from the import location. Given an economic cost of US$3.75/mmBTU for auto-diesel used at these plants (at the comparable US$17/bbl benchmark), the economic justification for conversions seems doubtful. Given other more pressing priorities for infrastructure investment, it is therefore hard to see an LNG power project serving as a credible catalyst for more widespread use of gas in Sri Lanka.

Project Location

The worldwide experience suggests that project economics are strongly influenced by close integration of LNG shipping terminals accommodating large LNG tankers, LNG storage, regasification plant, and immediately adjacent power projects.

Prima facie the optimal location for an LNG terminal in Sri Lanka (as indeed for a coal-based project) is Trincomalee, which benefits from one of the best natural harbours in the region, and is one that could readily accommodate Capesize vessels. However, given the security situation in the Trincomalee area (and, just as important, the international perceptions of the security situation), whatever may be the economics of LNG, such a project location is simply not bankable for the foreseeable future.

Colombo port is already quite congested, and it is hard to see how an integrated LNG project could be accommodated in the Port area. While the worldwide LNG industry has an impeccable safety record, such a location would carry very high insurance costs (if indeed bankable at all), given the history of terrorist attacks in and around Colombo (on oil storage depots and in the nearby downtown area itself). Other sites on the west coast (such as Puttalam) pose significant safety risks, because during the monsoon season wave heights constrain LNG tanker operations for safety reasons.

Project Economics

The economics of LNG projects thus depend largely upon scale, supply (and transportation) mode, and location. If we exclude pipeline supply from India and Bangladesh, there are three options.[16] In the first, larger (>100,000 m^3) LNG tankers serving south Indian projects from the Gulf would make their return journey via Sri Lanka, retaining in their holds such smaller quantities as may be required in Sri Lanka. But whether the LNG vessels used to supply large Indian terminals would be suitable for the relatively shallow waters off Puttalam is doubtful. Moreover, the long length of jetty required even to reach the minimum required water depth imposes additional costs and handling difficulties: the optimal configuration (as would be possible at Trincomalee, and as is practised at typical Japanese terminals) calls for a compact arrangement of unloading point, storage and regasification.

The second option is to use small LNG carriers (30,000 m^3) under direct charter, directly supplying a 2×300 MW scale project. But this is likely to be relatively expensive, lacking scale economies.

A third option involving backhaul was proposed by ENRON. Large LNG tankers from the Gulf presently supply Japan. On their return journey, normally empty, they would stop at an Indonesian LNG terminal (such as Arun at the tip of Sumatra), take on a partial shipment that is carried to Sri Lanka, unloaded at a southern port (such as Galle), and then, again empty, return back to the Gulf. Though the incremental travel distance to an Indonesian (or Malaysian) LNG terminal would be relatively small, additional loading and unloading times would add to costs. Nevertheless, the total shipment cost is almost certainly lower than those of the first two options, although how much of this saving would accrue to Sri Lanka rather than to the supplier is unclear: as a small buyer in a market dominated by large buyers in India, China and Japan, Sri Lanka will have limited negotiating power.

Even at the more common 2–2.5 mtpy project size, there are few detailed economic and financial analyses in the public domain. Moreover, developers' project proposals to Governments are understandably reluctant to include detailed financial spreadsheets – financial detail was completely absent from ENRON's proposal.

There are a few references in the literature to small LNG projects, and Table 4.5 shows an illustrative calculation for the gas price.[17] If the fob Gulf price were US$3.00/mmBTU (equivalent to a crude price of about around US$17/bbl), the burner tip price at Puttalam would be US$5.05/mmBTU. Even if we assume stable crude prices at US$11/bbl (an unlikely event), the delivered LNG price would still be at around US$ 4.05/mmBTU.[18]

Table 5.6 Price build-up for small LNG project in Sri Lanka (in US$/millionBTU)

		Sri Lanka	
	Hypothetical LNG-based IPP(a)	Oil price US$16/ bbl	Oil price US$11/ bbl
LNG price, loaded on ship (fob Gulf for Sri Lanka) 432,000 tonnes/year	3.00	3.00	2.00
Shipping in small 29,590 m^3 carrier (Century); 186,000 bbls/shipload, 1,600 km distance	0.59		
Shipping to Sri Lanka: 2,700 km: same LNG carrier		1.00	1.00
LNG terminal (assumed capital cost $80.8 million)	0.85	0.85	0.85
Pipeline to power plant	0.16	0.20	0.20
Delivered to plant	4.6	5.05	4.05

Source: Based on Mitchell (1997, p. 39), adjusted to Sri Lanka conditions

Economic Analysis

We have examined LNG in two different implementation modes. The first is based on the assumption of a 750 MW power-only project, for which we estimate the overnight capital cost (with penalty for up-front infrastructure) at US$1,460/kW (including terminal and regasification). If LNG fob is priced at US$2.81/mmBTU (corresponding to an Arab Light price of US$18.5/bbl), and an efficient backhaul transportation system keeps freight costs to US$0.20/mmBTU, then the cif Sri Lanka price would be US$3.00/mmBTU. However, as shown in Figure 5.4, at this price LNG would not be competitive with coal for base-load duty.

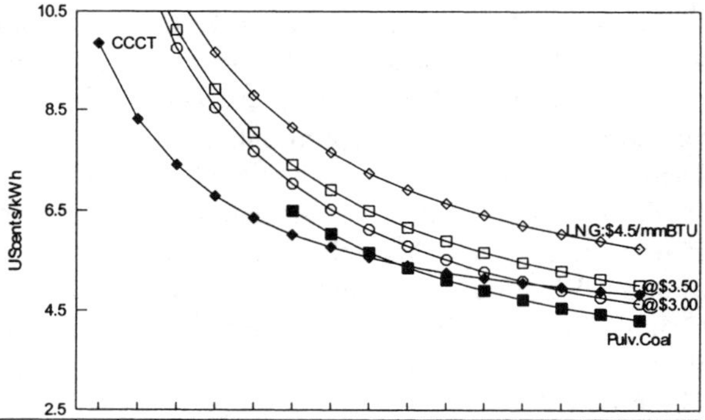

Annual PLF		0.2		0.3		0.4		0.5		0.6		0.7		0.8		0.9
■ Pulv.Coal						6.5	6.0	5.7	5.4	5.1	4.9	4.7	4.6	4.4	4.3	
◆ CCCT	9.9	8.3	7.4	6.8	6.4	6.0	5.8	5.6	5.4	5.3	5.2	5.1	5.0	4.9	4.8	
⊖ @$3.00	14.6	11.6	9.7	8.5	7.7	7.0	6.5	6.1	5.8	5.5	5.3	5.1	4.9	4.8	4.6	
⊟ @$3.50	14.9	11.9	10.1	8.9	8.1	7.4	6.9	6.5	6.2	5.9	5.7	5.5	5.3	5.2	5.0	
⊖ LNG:$4.5/mmBTU	15.7	12.7	10.9	9.7	8.8	8.2	7.7	7.3	6.9	6.7	6.4	6.2	6.1	5.9	5.8	

Figure 5.4 Screening curves for LNG (power-only project)

The corresponding life-cycle costs are indicated in Table 5.7. The probable minimum price of LNG corresponding to a world oil price of US$ 8.5/bbl is US$ 3.0/mmBTU, falling to US$ 2.75/mmBTU at US$ 16/bbl (for which the NPVs and penalties are typeset in italics). We note that cost differentials are significantly higher than the difference between oil and coal (Table 5.2).

Table 5.7 Lifecycle costs (US$ million, NPV at 10 per cent), 75 per cent PLF

	Arab Light =US$18.5/bbl		Arab light =US$16/bbl	
	NPV	δ (coal)	NPV	δ (coal)
Coal	794		776	
LNG@ US$2.5/mmBTU	846	52	846	70
2.75	881	87	*881*	*105*
3.0	916	122	916	140
3.5	986	192	986	210

The second mode of implementation is under the assumption that the project would be of much larger scale, with significant non-power off-take of gas (which, as noted previously, would entail a large additional infrastructure cost). Under such circumstances, the delivered gas price to the power project would be around US$ 3.75/mmBTU (though with a capital cost for the power project itself limited to the much lower CCCT price of US$ 725/kW).

Obviously, since the power generation costs themselves are the same as for autodiesel-fired CCCTs, and the diesel price is US$ 5.05mm/BTU (at the same assumed crude price of US$ 18.5/bbl), then any LNG price less than US$ 5.05/mmBTU would be competitive for intermediate duty. However, coal remains least-cost for baseload duty, a result that is consistent with Japanese experience.

These conclusions are also supported by a recent study of the potential for natural gas in Sri Lanka (Nextant, 2003) that takes a much more optimistic view of LNG (with an analysis based on a JCC crude price of US$ 18/bbl). It finds that a large terminal in India supplying gas to Sri Lanka would result in a price of US$ 5.50/mmBTU, substantially in excess of the US$ 4/mmBTU the study claims to be required to be competitive with coal; that gas pipelines from Bangladesh (and Andhra Pradesh) would be uneconomic;[19] and that gas demand is below the level required to make an LNG project economic, which the study estimates at 1.3 billion cubic metres (bcm), about a third of the 'traditional minimal size' of an LNG project of 3.5 bcm (2.5 mtpy). It argues that the cheapest option is a small dedicated terminal in Sri Lanka, that would result in a price of US$ 3.99/mmBTU (based on an fob Qatar price of US$ 2.46/mmBTU – substantially below current levels).

In short, LNG is not a viable power technology for Sri Lanka for the next decade. It is extremely unlikely that LNG would be offered in any IPP bid for either baseload or intermediate load duty. Moreover, as we discuss in Chapter 9, the arguments that the environmental benefits justify LNG, or that the incremental costs of LNG can be covered by the sale of carbon credits, are both weak: Sri Lanka has renewable energy options that can achieve carbon reduction at much lower costs.

5.4 CLEAN COAL TECHNOLOGIES

Atmospheric fluidised bed combustion (AFBC) is the clean technology that is closest to full commercial availability – though as noted above, its use is indicated mainly for poor quality coal (such as India's very high ash domestic coals). AFBC life-cycle costs are shown in Table 5.8. The magnitude of the incremental costs, compared to the baseline coal project, are significant, and even higher than coal with flue gas desulphurisation (FGD).

Table 5.8 AFBC Lifecycle costs (NPV, US$ million)

	NPV	δ (coal)
Coal	794	
Coal+FGD	887	94
AFBC	917	123

Note: Arab Light US$ 18.5/bbl; 75 per cent PLF, 10 per cent discount rate

Table 5.9 shows the corresponding sulphur emission levels, together with those of some of the other options examined in this volume. It is evident that for significant reductions in sulphur emissions, FGD would be the most cost-effective choice, given the use of high-quality Australian coal.

Application of clean coal technology does not appear to be warranted for Sri Lanka in the short- to medium term. The technologies that are at or close to commercial availability, such as AFBC, are most appropriate at mine-mouth plants using low quality coals (and hence their potential application in India and China). But because Sri Lanka has no indigenous fossil resources, and must therefore import high-quality coal over relatively long distances, the economic rationale for AFBC is absent.

Table 5.9 Summary of sulphur emission levels

	Mg/MJ	$KgSO_2$ /MWh	Tonnes/year for 300MW plant at 75%PLF
Proposed emission standard	520		
Coal, 0.5%S ('average' e.g. coal at Puttlam project)	372	3.57	7,030
Oil-steam, 0.8% sulphur	372	3.57	7,030
Oil-steam, 1.1% sulphur (a)	520	5.28	10,409
CCCT, auto diesel 0.1%S	45	0.36	706
Diesel using residual (b) at 1.75% S	813	6.64	13,087
at 3.5% S	1,626	13.28	26,174
Coal + FGD	38	0.38	740
AFBC	19	0.18	359

Notes:
S = sulphur
(a) This is equivalent to use of 3.5 per cent sulphur oil, treating 75 per cent of the stack gas by FGD @ 90 per cent removal.
(b) Though 3.5 per cent S is the specification, actual sulphur content of fuel oil used at Sapugaskanda is in the range of 1.75 to 2.25 per cent S. Particularly in recent years, this sulphur content has been lower than the sulphur specification, because a higher fraction of low-sulphur crude has been refined.

5.5 FINANCIAL ANALYSIS

A full analysis of the possible distortions resulting from customs duties and excise and other taxes is beyond the scope of this book. However the most important issue concerns the cost of capital, and the extent to which the financial cost of capital differs to the discount rate used for economic analysis. To illustrate the impact of private v. public financing, we make the following assumptions:[20]

- 30:70 debt equity ratio, with equity disbursements during the construction period *pari passu* with debt.
- For privately financed IPPs, 12-year offshore dollar financing at 10 per cent interest. IDC is assumed rolled into the debt (with construction financing also at 10 per cent interest).
- For concessionary (Japanese) financing we assume 30-year tenure with 10-year grace and interest at 1.8 per cent.
- For concessionary finance, the equity contribution (of CEB or of its unbundled successor) is costed at 16 per cent.
- The financial capital cost is defined as the present value (at the opportunity cost of capital, 10 per cent) of the stream of equity returns and debt service payments.
- IPPs soft costs are taken as 20 per cent of capital cost.
- The effective cost of capital is defined as that discount rate which makes the stream of equity returns and debt-service payments (interest and principal) equal to the completed cost.

Implications of Concessionary Financing

At first glance, concessionary funding at interest rates of less than 2 per cent (as has been available to the Sri Lanka power sector in the past from Japan) seems very attractive, though for CEB that is less true, since the Government on-lends concessionary loans to CEB at substantially higher interest rates in the 10-12 per cent range (see Box 5.1 for a description of the debate over on-lending rates in Nepal).

Box 5.1 On-lending rates in Nepal

JBIC has supported several major power projects in Nepal, including the Kuhlekhani storage project and the 144 MW Kali Gandaki project (co-financed with ADB). The practice has been for the concessionary loans to be on-lent by His Majesty's Government of Nepal (HMGN) to the Nepal Electricity Authority (NEA) at 10.25 per cent (though with a lower rate for rural electrification projects). The stated rationale is that this is warranted to cover the exchange rate risk, since even concessionary loans must (eventually) be repaid in hard currency. NEA has requested that this on-lending rate be lowered as a way of alleviating its financial position (caused in part by the very high costs of power from Nepal's two foreign-financed hydro IPPs, the 60 MW Khimti Khola and 36 MW Upper Bhote Koshi projects).

This rationale for on-lending arrangements can be illustrated by simple example of a 150MW project with capital costs of US$2,000/kW, under the assumption of 27.5 per cent NEA equity and 72.5 per cent JBIC financing at 1.5 per cent interest, with 30 year maturity including 7 years grace, on-lent by HMGN to NEA at 10.25 per cent. The figure shows the surplus to HMGN (as NPV at 10 per cent discount rate) – as a function of the average rate of depreciation of the Nepali rupee (NRs) to the Japanese yen (JPY). The rate of currency depreciation has to exceed an average of 8.25 per cent for the Government to carry a net exchange rate cost. If the average rate of depreciation is 6 per cent, then the surplus to HMGN (as NPV) is NRs 4billion (US$53million).

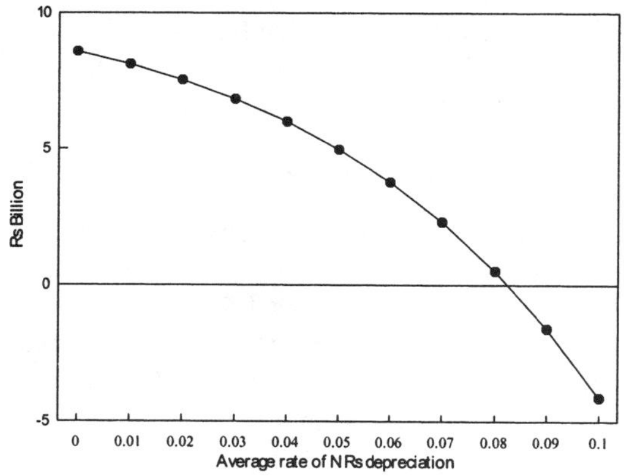

The actual historical rates of depreciation show significant variation. In the period 1972–85 the average annual rate was 7 per cent, followed by a decade (1986–95) of sharp depreciation (22.9 per cent), but since 1996 the average has been 3.1 per cent. Thus there have been periods when even a 10.25 per cent on-lending rate would still result in a net cost to HMGN. However, high rates of currency depreciation are generally associated with high rates of domestic inflation: therefore, if the exchange rate risk were passed to NEA, provided its tariffs were adjusted to the general inflation rate, the financial burden (in real terms) remains manageable. But with the Nepali Rupee firmly tied to the Indian Rupee, a return to the inflation (and currency depreciation) of the 1980s seems most unlikely. The Ministry of Finance is currently considering whether the on-lending rate to NEA could be lowered.

In Table 5.10 we show the difference between concessionary and IPP financing of the coal plant, calculated as the present value of the stream of equity returns and debt service payments. The results indicate very large differences, and certainly far greater than the differences between technologies noted in the economic analysis.

Table 5.10 Lifetime cost of capital (equity returns and debt service)

	Unit 1	Units 2 and 3	Average
Concessionary financing			
Completed cost	435	236	323
NPV	235	128	175
IPP financing			
Completed cost	650	315	431
NPV	823	446	611
Difference in NPV	588	318	436

Note: US$ million; NPVs at plant commissioning

Similar conclusions apply to other technologies as well. For example, the cost of hydro plants under concessionary financing is shown in Table 5.11. Again we see effective financial costs that are around 50 per cent of the assumed nominal capital costs. Other factors may raise financial costs. For example, the Upper Kotmale project has been repeatedly delayed by environmental protests. The increasing pressure to follow multi-stakeholder, multi-level consultations in order to achieve consensus is likely to increase costs further (see, for example, Munasinghe, 2004, for a discussion of both the advantages and pitfalls of such consultative processes).

Financing by the Japanese Bank for International Cooperation (JBIC) is necessarily denominated in yen, whereas IPP debt (particularly for large projects) is most likely to involve several participating banks and several currencies, and would most probably include a significant portion of debt denominated in US dollars. If the dollar continues to depreciate against the yen (since 1970 the historical average has been about 4 per cent per year, see Figure 5.5), then yen financing – even at concessionary rates – may be more expensive than the above comparisons suggest. This is especially true where long grace periods are involved: what effectively matters is the yen: dollar exchange rate 10–20 years hence.

Table 5.11 Effective cost of hydro plants

	MW	Overnight Cost		Penalty for uncertainty (as per CEB 1998 study)	Economic capital cost, CEB definition	Financial cost, JBIC financing
		US$	$/kW	(per cent)	$/kW	$/kW
Kukule	70	186				1,436
U. Kotmale	150	372				1,341
Broadlands	40		2,675	2	3,152	1,377
Ging Ganga	49		2,405	10	2,851	1,150
Moragolla	17		3,390	10	4,010	1,300
Uma Oya	150		2,324	5	2,876	1,163

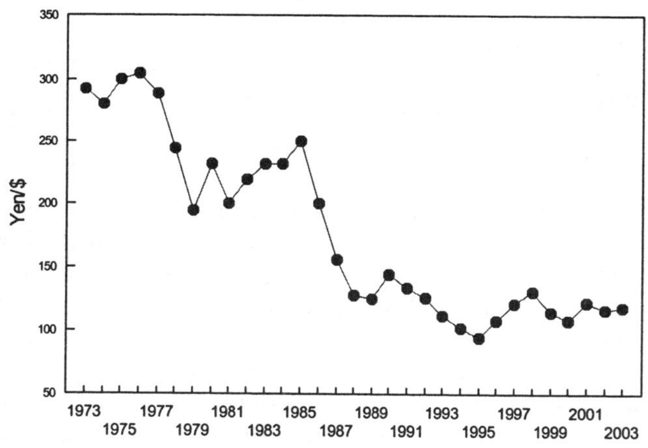

Figure 5.5: Yen to US$ exchange rate

The impact of long-term yen depreciation (Figure 5.5) on the financial comparisons is illustrated in Table 5.12. Even under the worst case (no yen, but 100 per cent US dollar financing in the IPP case, and an assumed 7 per cent year depreciation of the dollar against the yen – which exceeds the depreciation rate of any long-term historical period), the difference in NPVs for the first coal unit is still US$ 339 million.

Table 5.12 Impact of dollar depreciation: Financing of first coal unit

		Yen:$ year 10	NPV US$m	Equiv US$/kW	δ (IPP) US$m
IPP, 100 per cent dollar financing			823	2,742	
Concessionary financing, 100% Yen					
US$:yen depreciation rate	0%	100	235	784	588
	1%	90	255	849	568
	2%	80	278	926	545
	3%	72	306	1,019	517
	4%	65	339	1,130	484
	5%	58	379	1,262	444
	6%	53	427	1,422	396
	7%	48	484	1,614	339

The effective cost of capital associated with concessionary Japanese financing in yen is thus dependent upon the assumed depreciation rate of the US dollar against the yen. Given historical depreciation rates, and the probability that the corresponding privately financed IPP would also have some yen debt, it would seem reasonable to assume an effective cost of capital of around 7 per cent (rather than the nominal interest rates of less than 2 per cent).

However, it is worth noting that one of the consequences of the needless prevarication over the Puttalam coal project is that the earlier willingness of the JBIC to finance the project at concessionary rates has now been missed.

Cost of Capital of Privately Funded Projects

The financial cost of capital is determined by its constituent parts: the cost of the equity component, and the cost of debt. Both are best illustrated by example, for which purpose we use the oil-steam cycle plant proposed by Electricity Corporation of New Zealand (ECNZ), and for which ECNZ had proposed a two-part tariff: a fixed charge (to recover debt and equity), and a variable charge to recover operating and fuel costs. The fixed charge (which included both equity returns and debt service) first proposed in December 1996 was US$ 37 million. In the offer of October 1998 this was increased to US$ 47 million. Subsequently, it was reduced to US$ 41.5 million (further details of these offers are analysed in section 8 of Meier and Munasinghe, 2003c).

Box 5.2 Grant finance in Nepal

Nepal's power sector has been highly dependent on concessionary finance: indeed the power sector accounts for about 30 per cent of all donor assistance to Nepal, a significant portion of which is in the form of grants, though the proportion of grants to the power sector has fallen somewhat over the past decade.

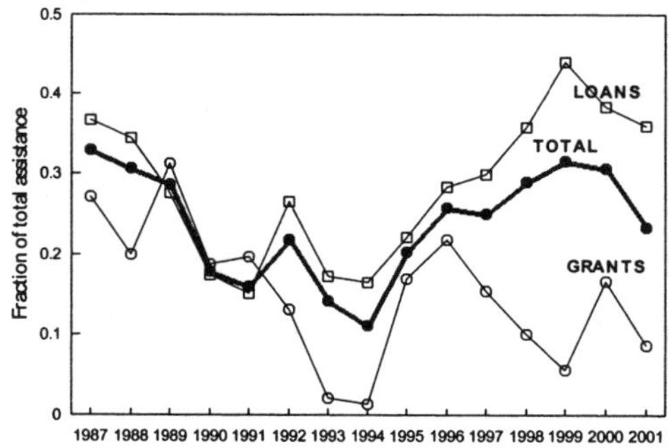

In Nepal, the practice of on-lending even extends to grants – where the exchange rate argument (see Box 5.1) cannot be invoked (because there is no repayment obligation). However, for NEA, grants are a poor proposition because equipment procurement under grants is almost always tied – which means substantially higher costs because the number of vendors is small and limited to the donor country.

For example, the high cost of NEA's multi-fuel diesel station supported by a grant from FINIDA (US$1,350/kW) is testimony to this problem. NEA estimates that, on average, such grant procurement leads to prices that are 30 per cent than under proper international competitive bidding (ICB).

Thus, when FINNIDA makes a grant of a diesel station that is supplied by Wartsila, say in the amount of US$20million, the losers and winners are as shown in the table. Here we assume that the grant arrangement results in a 30 per cent increase in cost, which means a surplus to Wartsila of US$ 5million (above the market price). The grant directly benefits HMGN. But the surplus that accrues to Wartsila is actually provided by higher costs to NEA's consumers, because the economic value (as revealed by ICB) is only US$ 15 million, and NEA must pay HMGN for a US$ 20million loan.

In US$ million	Donor Govt.	Wartsila	NEA	HMGN
Grant	-20	+20		
Economic cost of the diesel plant		-15		
Economic benefits			15	
NEA loan transaction			-20	+20
Net	-20	+5	-5	+20

Table 5.13 shows the relationship between the size of the fixed charge and the effective cost of capital, together with the NPV of the fixed plant charge – which is equal to Sri Lanka's effective capital cost (both financial and economic). We note that the 'revised offer' of US$ 41.5 million/year corresponds to an actual capital cost to Sri Lanka of US$ 353 million, as opposed to the nominal (completed) capital cost quoted by ECNZ of US$ 195 million.

Table 5.13 Cost of capital v. fixed charge

	Fixed charge (US$m/year)	IRR to ECNZ (%)	Cost of capital to Sri Lanka	Effective capital cost NPV $USm	Effective capital cost US$/kW
ECNZ offer of October 1998	47	33	23.6	400	1 802
ECNZ 'revised offer'	41.5	27.8	20.7	353	1 592
ECNZ offer of December 1996	37	23.4	18.2	315	1 420
Indian IRR benchmark	34	20.6	16.6	290	1 308
	30	16.5	14.2	255	1 151
	24	10.4	10.6	204	921
Fixed charge set equal to that value that yields a 10% cost of capital to Sri Lanka	23	9.4	10.0	196	882

Note: IRR= internal rate of return

It is evident from these results that, for the cost of capital to Sri Lanka to be no more than the Indian benchmark, which allows a generous 20.6 per cent IRR to the equity investor, the fixed charge should be no more than US$ 34 million (in the case of this particular project). Even the revised offer of US$41.5 million/year would have been a very unfavourable proposition for Sri Lanka, and implies an IRR to ECNZ of 28 per cent that is not only very high by current south Asian benchmarks, but is surely not commensurate with the small risks to the equity holders, given that ECNZ had asked for a full and unconditional sovereign guarantee.

More importantly, we see very large variations in effective capital cost, ranging from US$ 400 million in the case of ECNZ's October 1998 offer, to US$255 million at a more reasonable 16.5 per cent IRR, a difference of US$ 145 million. This is a comparable order of magnitude to the differences between private and concessionary financing, and points to the importance of competitive bidding for IPP projects.

When the earlier economic analysis is repeated at the 7 per cent (concessionary) financial cost of capital, the advantage of coal over oil-steam (other things like oil price equal) is more pronounced, as indicated in Table 5.14. Indeed, the lifetime financial costs show that even at US$ 13/bbl, coal still has an advantage over oil-steam.

Table 5.14 Lifetime costs at 7 per cent cost of capital (JBIC financing)

	US$13/bbl		US$18.5/bbl	
	NPV	δ (coal)	NPV	δ (coal)
Coal	855		906	
CCCT	947	92	1,224	317
Diesel (residual)	902	40	1,008	102
Diesel (furnace oil)	1,018	162	1,124	217
Oil-steam	866	10	1,008	102

Conclusions on Financial Analysis

While concessionary financing at interest rates of 1–2 per cent appears very attractive, the effective cost of capital is likely to be higher as a consequence of the probable continued depreciation of the US dollar against the yen. Depending upon assumptions, the actual effective cost of capital may be several percentage points higher.

Nevertheless, the effective capital cost (defined as the present value of the stream of equity returns and debt service payments) of JBIC financing is still substantially lower than an IPP financing. Even if the US dollar: yen depreciation rate is taken as an average of 4 per cent, and the IPP financing contains no yen debt component, the difference is in excess of US$300 million. Differences of this order of magnitude are significantly greater than differences between technologies.

At the lower costs of capital implied by concessionary financing, the more capital-intensive technologies will have an advantage. In the case of coal v. oil-steam for baseload, under JBIC financing coal has an advantage even at oil prices of as low as US$13/bbl (under baseline assumptions for capital cost). It seems reasonable to conclude that if JBIC financing is accepted, coal is a robust technology/fuel choice (though even at the cost of private capital, coal remains the fuel of choice for baseload for all but the very lowest oil prices not likely to persist for more an a few months (as occurred in late 1998).

The analysis of the fixed charges for the proposed Marsden B project in New Zealand shows clearly the importance of competitive bidding. Very high equity returns had been achieved by some IPPs in south Asia in the early years of opening the power sector to IPPs, when a negotiated Memorandum of Understanding (MoU), rather than competitive bidding, was the fashion. Such high equity returns may well have been justified by the risks involved (especially in India), but even risk-adjusted returns are best determined by competition.

Recitations of capital cost by IPPs give a misleading picture of the true cost of capital, for the actual cost of capital investment will be given by the present value of the stream of fixed plant charges (equity returns and debt service) – which as seen in the case of the latest ECNZ offer, amount to US$ 353 million – compared to the US$ 195 million stated to be the completed capital cost. We note that such differences are again not only greater than the differences between technologies (for equivalent duty), but comparable to the differences between private and concessionary financing.

The differences in cost attributable to differences in debt financing terms are also far smaller than the range of costs implied by the variations in equity returns encountered in MoU negotiations. Even if JBIC funding for power projects is tied (that is, not immediately available to other sectors such as health and education, were the power project to be funded by some other means), for economic analysis the opportunity cost of capital should still be taken at the SDR (10 per cent as assumed here) even though the financial costs are significantly less.

One of the arguments against concessionary financing is that this creates an additional actual or contingent liability upon the Government for which there may be limited headroom. On the other hand, if sovereign guarantees are offered to IPPs, the Government incurs an even larger contingent liability (since the amounts to be guaranteed to IPPs are very much larger than in the concessionary financing case because the cost of capital is also higher).

However, this is a transition issue. Sovereign guarantees may need to be offered to the first few IPPs, but subsequent projects may be able to proceed without them (provided power sector reform as a whole proceeds satisfactorily, and that the regulatory commission establishes a track record of adequate tariff-setting). Indeed, this is the experience in India, where the first few IPPs (the seven so-called 'fast track' projects, including ENRON's Dabhol project) received federal guarantees, but none are to be given in the future.

5.6 THE PRESENT SITUATION

In mid-2004, the price of Australian coal fob NSW is US$60/tonne (compared to just US$24/tonne in June 2003), which translates to a coal price cif Sri Lanka of around US$ 72/tonne (compared to US$50/tonne in 1999-2000). Brent trades in the US$ 35–40/bbl range. There is much talk of a decoupling of oil and gas prices, and the pricing of LNG delivered to Japan is becoming more flexible, particularly with the entry of China as a major LNG importer. In 2002 and 2003, Indonesia and Australia are reported to have contracted to supply gas to China's Fujian and Guangdong import terminals at US$ 2.40–3.00/mmBTU (at a time when Japanese consumers were paying US$ 4.30/mmBTU).[21] Nevertheless, for Sri Lanka, the problems of small scale remain, and in April 2004 the actual average LNG price cif price to Japan was US$ 4.90/mmBTU.

Figure 5.6 shows the screening curve under current conditions, based on US$ 72/tonne coal (cif Sri Lanka), a Brent price of US$ 35/bbl (which translates to around US$ 46/bbl for auto-diesel cif Colombo), and a cif LNG price of US$ 4.90/mmBTU. Pulverised coal, even at US$ 72/tonne, retains its advantage.

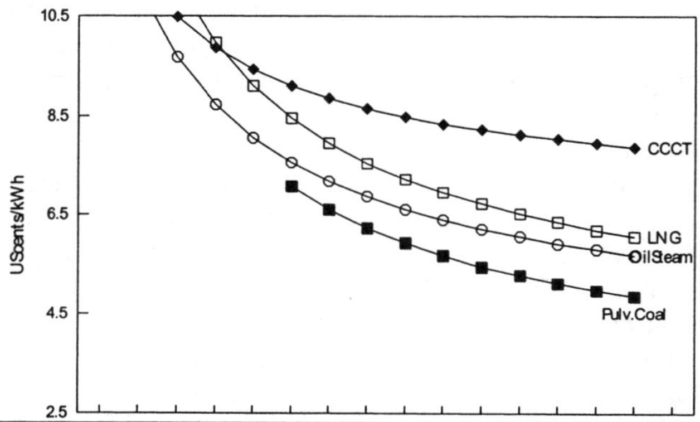

Annual PLF		0.2		0.3		0.4		0.5		0.6		0.7		0.8		0.9
■ Pulv.Coal						7.1	6.6	6.2	5.9	5.7	5.5	5.3	5.1	5.0	4.9	
◆ CCCT	12.9	11.4	10.5	9.9	9.4	9.1	8.9	8.6	8.5	8.3	8.2	8.1	8.0	8.0	7.9	
⊖ OilSteam	13.4	11.1	9.7	8.7	8.1	7.6	7.2	6.9	6.6	6.4	6.2	6.1	5.9	5.8	5.7	
⊟ LNG	16.0	13.0	11.2	10.0	9.1	8.5	8.0	7.6	7.2	7.0	6.7	6.5	6.4	6.2	6.1	

Figure 5.6 Screening curve, mid 2004 price levels

5.7 CONCLUSIONS

For baseload duty, a steam cycle plant is indicated. While coal is least cost under most sets of assumptions, the switching value analysis suggests that steam-cycle oil plants, burning imported fuel oil (and blended to meet the proposed sulphur oxide emission standard) may be competitive at oil prices (Brent) consistently below US$ 16/bbl. Given that such an oil price is extremely unlikely in the foreseeable future, except possibly as a temporary cyclical phenomenon (similar to the oil price decline in 1998), we can say with some confidence that there continues to be no reason why Sri Lanka should not implement a baseload coal plant as soon as possible.

Just as clear is that combined cycle plants running at load factors exceeding 55–60 per cent are not viable for baseload duty. Combined cycle plants burning auto-diesel or naphtha are indicated only for intermediate duty (and to cover peak loads in dry years). If any additional combustion turbines are built in Sri Lanka (beyond the AES project), they should be naphtha-powered (at least to the point at which, in average hydro years, naphtha no longer needs to be exported).

LNG is not a viable option for Sri Lanka in the next decade. Even if the Puttalam baseload project were bid as an IPP and the fuel choice were left to the IPP to decide, it is extremely unlikely that any IPP would bid an LNG project (quite apart from the question of whether LNG could be unloaded at this exposed site). If a combined cycle project for the South Coast (say at or near Galle Harbour) were advanced, again it is extremely unlikely that an LNG-based project would be among the top-ranked proposals.

Simple one-on-one comparisons (and screening curves) appear to give results that are consistent with those given by more sophisticated least-cost system planning tools, and serve a valuable function particularly for sensitivity analysis which is more easily conducted in such simpler models. They also have the merit of transparency.

But these simple tools also have important limitations. They cannot say anything about when investment decisions should be made; about the optimal size of plants required to meet given load increments; or about the mix of peaking, intermediate and baseload plants that are required to meet given levels of reliability. Only more sophisticated systems planning models are suitable to answer these latter questions (although as we note below, even system planning models should not be used as the sole basis for decisions, and need to be supplemented with other decision-analysis tools). The systems planning approach is therefore used in Chapter 9, where we examine options for GHG mitigation in the power sector.

The differences in lifecycle costs among viable technology options for the same duty (for example, coal or oil-steam for baseload) are significantly smaller than the differences due to choice of financing (IPP or concessionary financing), or the costs of delay. It necessarily follows that the choice between coal and oil as a fuel is far less important than the choice of IPP or concessionary financing, or making decisions in a timely way.

At the same time, without a more formal decision analysis (see Chapter 10), it is difficult to assess the real impact of uncertainty. The conclusions about the need for a baseload plant in 2004 are critically dependent upon assumptions about load growth, or, for example, whether the generation requirement can be reduced by T&D loss reduction or demand-side management (DSM). To be sure, there may be good grounds, given the inevitable delays in implementing projects, to err on the side of caution, especially when the costs of inadequate capacity during dry years may impose large economic costs (as occurred in 1996). However, one might more reasonably conclude that it is better to improve decision-making and project management through reform and restructuring.

Indeed it is clear that the capacity shortages in the 1996 drought year were a consequence of the failure to make timely decisions for additional capacity, not poor forecasts of either demand or hydrology: the same is true of the more recent shortages experienced in 2001–04. That oil prices have risen to levels that would have seemed unreasonably high in 1999–2000 changes little: there simply is no reasonable alternative to a coal-fired baseload plant, fitted with modern environmental controls, and whose actual environmental impacts will be much smaller than oil-based projects near population centres.

In short, the proposition that auto-diesel CCCTs (or even residual oil steam-cycle plants) represent a viable and sustainable alternative to coal for baseload duty is simply incorrect; the proposition that small diesel units (much favoured by local private entrepreneurs) represent a alternative is demonstrably absurd; and the proposition that natural gas will be found in Sri Lanka's (or even south Indian) offshore waters, or that the price of LNG will become competitive with coal for power generation, is wishful thinking.

Above all, now as in 1998–2000, the main problem is Government prevarication and paralysis in taking decisions (and subsequently reversing them for purely political reasons). Given the polarisation of politics and razor-thin parliamentary majorities, a situation unlikely to change in the near future, this paralysis may be unavoidable for as long as the power sector remains in State ownership and control. Achieving real independence from Government interference must therefore be one of the main aims of the power sector reform and restructuring process.

NOTES

1 Needless to say, the question of the discount rate is not without controversy, particularly in the case of economic analysis of climate change issues where time horizons are very long. This is discussed further in Chapter 9.

2 Some of the more important discrepancies concern assumptions for capital and non-fuel operating costs for oil steam-cycle plants.

3 This is the assumption used by ECNZ in their proposal for a 250 MW oil steam-cycle plant. However, if the emissions standards as proposed by the Central Environment Authority were actually implemented, the sulphur content of heavy fuel oil would need to be reduced to around 1.1 per cent S.

4 For example, the overnight capital cost of coal is taken as US$ 1,008/kW (as per Electrowatt Engineering, 1996, Table 3.5). This is adjusted to 1999 prices to yield US$ 1,071/kW. This is adjusted by 7 per cent for the up-front-infrastructure penalty of 7 per cent to yield US$ 1,146/kW. Finally this is adjusted for the opportunity cost of capital to US$ 1,303.

5 The difference between an outlay of λ, *n* years hence, with an outlay of λ today, is $\lambda - \lambda/(1+r)n$ where *r* is the discount rate.

6 The question of carbon offsets, and the related policy options available to the Government of Sri Lanka, is discussed in more detail in Meier and Munasinghe (2003c).

7 In the case of the Kerawalapitiya site, shallow water conditions would require a pier distance of 4.6 km to accommodate tankers. Moreover, safety standards require maximum wind velocity of no more than 15 metres/sec, and maximum wave height of 0.5 metres, which in the May-September monsoon period is satisfied only 3 days/month. It also seems unlikely that the wave and wind conditions at the Puttalam site would be substantially better than at Kerawalapitiya. Finally, we note that the safety standards cited are the Japanese Technical Standards for Port and Harbor facilities: application of lesser standards in Sri Lanka would unlikely be acceptable.

8 A comparison of the 1996 and 1998 CEB generation plan assumptions is instructive. In 1996 the capital costs of open cycle gas turbines were estimated at US$ 514/kW; in 1998, this had dropped to US$ 372/kW, an assumption documented by the actual price of a 105 MW unit commissioned in 1997. Nevertheless it is worth noting that in 1988, the Thermal Options study (Black and Veatch, 1988), estimated capital costs of a 300 MW coal plant at around US$ 1,600/kW, which a decade later had fallen to US$ 1,451 (for the first unit of the Puttalam project), equal to about US$ 1,000/kW in 1988 prices, a decrease of 40 per cent in real terms. Both estimates include the cost of coal jetties.

9 Marketing rhetoric abounds: 'LNG strategy should be driven by the electricity and gas market, not by the requirements of an LNG producer . . . being independent from LNG suppliers has allowed ENRON to do what is best for the project and the downstream users.'

10 To be sure, one may always say that the cancellation of the ENRON project by the incoming State Government in 1995, and its subsequent reinstatement after protracted renegotiation (and granting of an equity share to the Maharashtra State Electricity Board), constituted an unusual circumstance that would not be encountered in future projects. However it must be understood that the main constraint to a smooth implementation of Dabhol has always been the lack of transparency, and high costs, associated with the MoU arrangement. Such negotiated MoU projects – especially where the cost of power is substantially above that of its alternatives – will always be publicly controversial, and involve high risk of delay.

11 Proposed projects in Gujarat contemplate injecting LNG into the existing Hazira-Bijaipur-Jagadishpur (HBJ) pipeline, which was originally constructed to bring India's

domestic off-shore gas reserves to the industrial regions of Gujarat and the Delhi metropolitan area.

12 Assuming a cost of US$ 0.625-1.25 million per km, and a distance of 560 km (to a terminal at Cochin), the investment cost would be US$350-700 million, plus terminal and storage facilities at either end. This would add at least US$ 2/mmBTU (to the Indian terminal regasified LNG price of around US$ 3.75/mmBTU) price at the sort of volume required in Sri Lanka even under any maximum utilisation scenario. In short, if indeed LNG were to be used in Sri Lanka, there is no reason why the LNG terminal would not be in Sri Lanka itself, rather than supplied by pipeline from India. The only reasonable scenario under which piped gas supply from India might ever be possible is if there were some large gas find in (or offshore) southern India itself. But there are no indications that such reserves exist: the most recent gas finds in the Krishna-Godaveri basin are much further north in Andhra Pradesh.

13 However, ENRON declined to provide CEB with the details of the burner-tip gas price at the Puerto Rico plant (EcoElectrica), or of the electricity selling price.

14 Even in the dry zone, or in southern coastal locations with drinking water problems (where coastal aquifers suffer from depletion and saltwater intrusion), the cost of desalinisation far exceeds the cost of bringing water by pipeline from the wet zone.

15 1998 consumption of diesel was about 1.06 mtpy, and gasoline 0.2 mtpy.

16 Generally, gas pipelines are cheaper than LNG at distances below 3,000 km.

17 As detailed in Mitchell (1997, p. 39). We may note that Mitchell's analysis of such an LNG-based 400 MW IPP results in a price of US 5.29 cents/kWh (as compared to the US 3.86 cents/kWh estimated by CEB for the Puttalam coal plant.

18 For a good review of the LNG market in Asia, see Petroleum Economist, Sept. 1998, p. 78–80. Thailand has recently postponed its planned LNG terminal, and the Korea Gas Company and other major LNG importers are attempting to renegotiate contract deals. Only baseload contracts are being taken up. This has caused concern for some of the big new LNG export projects in the Gulf; several have already been delayed (and some cancelled). Indeed, it is obvious that if the crude price stays below $15/bbl for any extended period of time, many new upstream LNG projects are unbankable. The TERI study of India notes that development of new projects in Indonesia (such as Natuna) will require prices of over US$ 4.25/mmBTU. Nevertheless, whatever may be the price of LNG, when LNG prices are low, fuel oil prices will be even lower.

19 A pipeline from Bangladesh would be over 2,000 km in length, and would need to traverse difficult marine environments (for example, the Ganges/Jamuna fan where there is significant sea-floor instability, see Nexant, 2003). A gas pipeline might also be possible from the Krishna-Godaveri Basin (about 1,200 km in length), but there are doubts as to whether the newly discovered reserves would allow a major export project.

20 These assumptions are discussed further in section 8.1 of Meier and Munasinghe (2003a). The detailed calculations for each case are included in Annex VI of that report.

21 One of the important developments over the past few years has been a significant decrease in transport costs, partially offsetting higher fob prices that are tied to crude prices under the traditional formula for LNG pricing to Japan.

6. The economic and environmental costs of traffic congestion*

Urban traffic congestion, particularly in the rapidly growing major metropolitan areas of developing countries, is imposing increasing economic and environmental costs. Traditional methods of curtailing congestion have largely failed, particularly on the supply side: reduction of congestion following expansion and improvements of roads is typically short-lived,[1] and further constrained by increasingly tight fiscal, physical and environmental constraints. However, demand side options, particularly pricing,[2] offer a more promising long-term solution – as evidenced by the successful pricing scheme in Singapore,[3] one of the few large Asian cities free of major road congestion.[4]

The bulk of the traffic entering the city of Colombo enters on seven main radial roads, which all connect to the central business district (CBD). These roads all exhibit very strong peak-hour traffic: Figure 6.1 shows typical inbound and outbound patterns for the A3, and Figure 6.2 shows comparisons of the outbound traffic counts for three of the main radial roads.

Is there traffic congestion in Colombo? Whatever the anecdotal evidence, there is a rigorous definition: congestion can be said to occur at the point at which an additional vehicle entering a road segment causes the normal speed of other vehicles to decrease, thereby imposing a cost not only on that additional vehicle, but on all vehicles using that road segment (Jayaweera, 1999). Based on a definition of congestion as occurring whenever average speeds of 30 km/hr cannot be maintained, the Colombo Urban Transport Study (CUTS) (W. S. Atkins and University of Moratuwa, 1995a) estimates the costs of congestion at Rs 11 billion/year.[5] This is clearly significant and which, in the absence of policy reforms, will inexorably rise. Already average speeds in the Greater Colombo Metropolitan Region (GCMR) have dropped to about 22 km/hour, and range from 10–15 km/hour in the centre to around 45 km/hour at the periphery of the GCMR.

* Chapter 6 contributors: D. Jayaweera, A.S. Kumarage, S. Chandrasiri, and R.M. Munasinghe

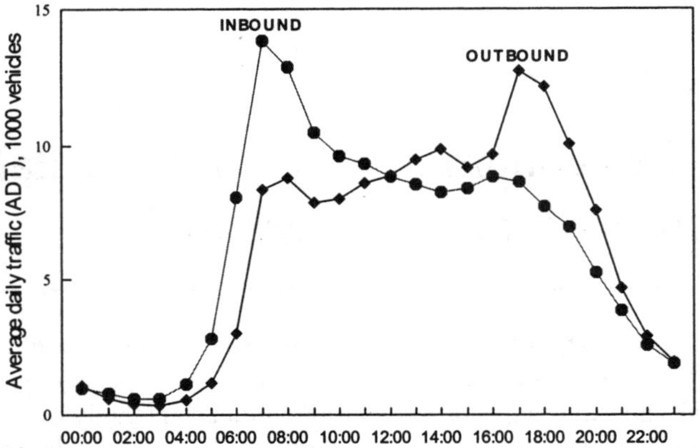

Source: Traffic Data from the traffic and planning division of the RDA

Figure 6.1 Hourly average daily traffic by hour for the A3 (Peliyagoda–Puttalam Road)

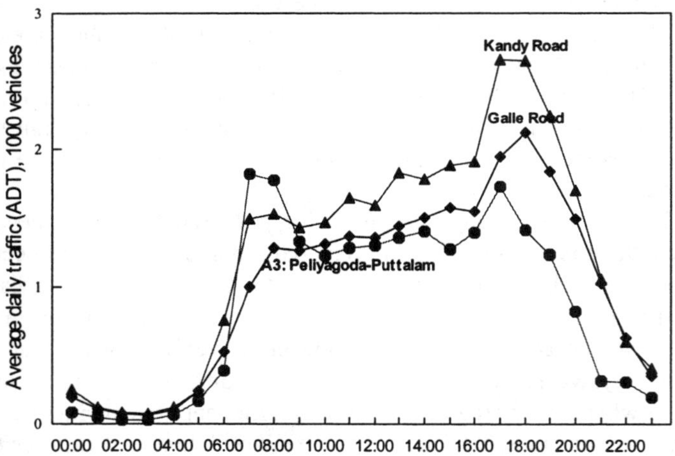

Figure 6.2 Traffic patterns on radial roads (outbound direction)

The rapid decline of average speeds on many of Colombo's radial roads over the past few years has already been noted (Table 1.1). Over the past four years, speeds on all roads have fallen by at least half; moreover, while in 1997 peak conditions were observed only for roughly two hours during the morning peak, by 2001 these conditions extended until mid-afternoon (in the inward direction); and similarly in the outward direction, the congested period now starts already in mid-afternoon. Such dramatic reductions in network speeds result in significant increases in local air emissions, because fuel efficiency decreases non-linearly with speed – especially under stop–go driving conditions of highly congested roads. This is documented by survey data taken in March 2001 on the Colombo–Ratnapura road: as shown in Figure 6.3, particularly for buses and trucks fuel efficiency falls dramatically at low average speeds.

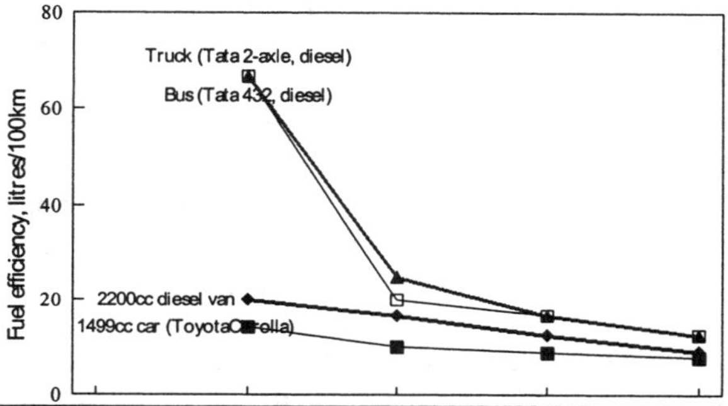

speed, km/hour	10	20	30	40
■ 1499cc car (ToyotaCorolla)	14.3	10.0	8.7	7.7
◆ 2200cc diesel van	20.0	16.7	12.5	9.1
▲ Truck (Tata 2-axle, diesel)	66.7	25.0	16.7	12.5
⊟ Bus (Tata 432, diesel)	66.7	20.0	16.7	12.5

Source: Fuel efficiency survey, March 2001, Ministry of Transport

Figure 6.3 Fuel efficiency survey for typical vehicles

A microeconomic analysis of congestion pricing in Colombo has been undertaken by Jayaweera (1999). The study derived the optimal value of a congestion toll by setting it equal to the difference between marginal social cost and average private cost. In this chapter we extend this micro-economic work to a calculation of the aggregate economic effects, and to an estimation of the related environmental benefits. However, we begin with a review of price elasticities.

6.1 PRICE ELASTICITIES

In this section we examine the influence of prices on decisions of road users. A first set of questions concerns choices made at the time of vehicle purchase (and particularly on whether to buy a diesel or gasoline powered vehicle) – likely to be a function of consumers' expectations about the cost of operation (a matter, in turn, of expectations about the future price of different fuels at the pump), and first cost. The second set of questions concerns the aggregate response of all road users to changes in fuel prices at the pump. Rates of taxation play as much a role in shaping these decisions as world oil prices and cif prices of automobiles: rates of customs and excise duty on vehicle imports show wide variations across different vehicle types, made more complex by provisions for exemptions and concessions to certain types of owners (which are frequently changed).[6]

The demand for vehicles in Sri Lanka has been studied by Chandrasiri (1999). Vehicles are not manufactured in Sri Lanka,[7] so the entire demand is met by imports, of which only some 25 per cent are new vehicles. Today the typical passenger car import is a reconditioned Japanese vehicle 2–3 years old. Unfortunately, the available data are of poor quality, because such records as are kept by Sri Lanka customs do not have the requisite detail on engine capacity, year of manufacture, etc.

In Figure 6.4 we show the average nominal price, ex-customs,[8] for the four main types of passenger vehicle, expressed in US dollars: most of the price variation reflects changes in taxation.[9] The sharp decline in purchase prices in 1978 reflects the lowering of import duties at the time of liberalization of the economy.

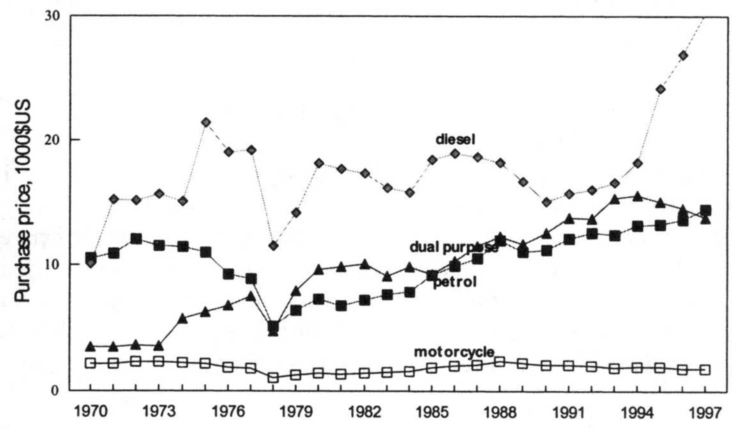

Figure 6.4 Vehicle prices, in US$1000

Figure 6.5 shows the corresponding data in constant 1982 rupees. The increase in the price of diesel vehicles since 1995 – a reflection of the imposition of so-called 'luxury taxes' – is largely cosmetic, for the number of diesel passenger cars actually sold is quite small (most of the growth being in 'dual-purpose' vehicles, for even in Sri Lanka SUVs have become fashionable displays of status),[10] and a significant proportion of those that are imported have benefited from special permits and exemptions.

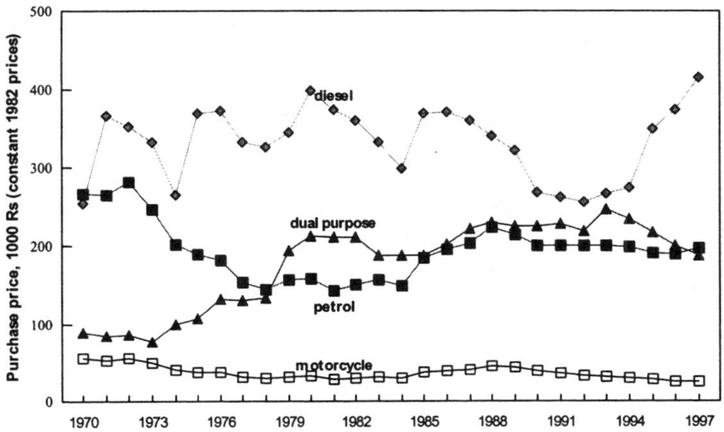

Figure 6.5 Vehicle prices, in constant 1982 rupees

In Table 6.1 we show a comparison of the taxes levied on diesel and petrol cars, using the same Toyota model differing only by type of engine. The import prices are for reconditioned, 3-year old vehicles. The (absolute) tax burden for the diesel vehicle at the time of purchase is seen to be double that of the gasoline-powered car, though as percentages of the initial cost the tax burden is comparable.

Table 6.1 Comparison of the diesel and petrol vehicle price for the Toyota model RAV4(a):costs at purchase

Cost in rupees	Gasoline	Diesel
Costs at purchase		
CIF	1,800,000	2,200,000
Import duty	267,000	534,000
Defence levy	108,489	217,000
Excise duty	254,400	508,800
GST	198,721	397,400
Stamp duty	21,380	42,800
Total tax	850,092	1,700,000
Total CIF+tax	1,650,092	3,900,000
Tax as % of purchase price	52%	44%
Registration	3,500	6,500

Diesel tax		10,000
Luxury tax (a)		50,000
Annual fuel cost		
km/year	15,000	15,000
km/litre	10	10
litres/year	1,500	1,500
Price/litre	50	21.50
Annual fuel cost	75,000	32,250

Note: (a) Luxury tax declines by Rs 10,000/year, reaching zero in the 6th year. Figures are tabulated for taxes and typical prices as of April 2001.

Several estimates of price elasticities of aggregate transportation fuel demand in Sri Lanka are available in the literature (Table 6.2).[11] However, many of the earlier estimates are limited to the period 1970–85, and do not properly capture the major changes in road transport consequent to liberalization of the economy in 1977.[12]

Table 6.2 Estimates of own-price elasticity

		Gasoline	Diesel
Sri Lanka			
de Silva (1989)	1970–85	-0.55	-1.21
Samaraweera (1989)	1970–85	-0.387	-0.26
Ranasinghe (1989)	1970–85	-0.508	-0.034
Meier et al. (1993)	1973–91	-1.14 (long-run) -0.334 (short-run)	
McRae (1994)	1973–87	-0.337	
Jayaweera (1999)		-0.163 (long run) -0.115 (short run)	-0.339 (long run) -0.154 (short run)
Other countries			
Philippines (McRae, 1994)	1973–87	-0.391	
Thailand (McRae, 1994)	1973–87	-0.304	
Malaysia (McRae, 1994)	1973–87	-0.125	
S.Korea (McRae, 1994)	1973–87	-0.496	
S. Korea (McRae, 1994)	1973–97	-0.866 (long-run)* -0.385 (short-run)*	
Indonesia (McRae, 1994)	1973–87	-0.197	
India (McRae, 1994)	1973–87	-0.321	
Taiwan (McRae, 1994)	1973–87	+0.024	
Taiwan (Garbacz, 1989)	1954–85	-1.362(long-run) -0.245(short run)	
Taiwan (Banaszak et al, 1999)	1973–92	-0.519(long-run)* -0.124 (short-run)*	

Note: (*) Demand for gasoline and diesel taken together (see text discussion)

Specification problems abound. In the case of gasoline demand studies, definition of the independent variable as consumption per vehicle runs into the problem of rapid changes in the vehicle mix, notably the significant growth in motorcycles (whose fuel demand per vehicle is a small fraction of that of automobiles).[13] Some studies add a variable for vehicles per capita – whose elasticity is almost always strongly negative:[14] for example, in McRae's Sri Lanka estimates, the demand elasticity with respect to vehicle/capita has a value of -0.504. Among past Sri Lanka studies, only Meier et al. (1997) includes a variable for fraction of motor cycles in the vehicle mix (with a corresponding elasticity of -0.17).

Among the various studies of Asian countries in the literature, the most rigorously specified is that of Garbacz (1989) for Taiwan, with who estimated the model

$$AUTO = a_1 PGAS^{a_2}\, PBUS^{a_3}\, INCOME^{a_4}\, AUTO^{a_5}\, MCYCLE^{a_6}$$

$$MCYCLE = b_1 PGAS^{b_2}\, PBUS^{b_3}\, INCOME^{b_4}\, AUTO^{b_5}\, MCYCLE^{b_6}$$

$$GAS(t) = c_1 GAS(t-1)^{c_2}\, PGAS^{c_3}\, INCOME^{c_4}\, AUTO^{c_5}\, MCYCLE^{c_6}\, TAXI^{c_7}$$

where *AUTO* = number of automobiles per capita
PGAS = gasoline price (per litre)
PBUS = bus price per passenger km
INCOME = personal disposable income per capita
MCYCLE = number of motorcycles per capita
GAS(t) = gasoline consumption per capita at time t
TAXI = number of taxicabs per capita

However, even this study was unable to include vehicle price variables – though it is worth noting that this study of Taiwan, and the Sri Lanka study by Meier et al. (1997), show similar results for both long- and short-run elasticities (see Table 6.2), and both are consistent with the value of -1.05 for long-run elasticity reported by Dahl and Sterner (1991) as the average value across developing countries.

Since gasoline and diesel are partial substitutes, rigorous estimation of price effects need to take into account the relevant cross-price elasticities. Therefore price and income elasticities must be estimated simultaneously. There are two ways of doing this: (1) combine gasoline and diesel consumption, and use a price variable that is the weighted diesel and gasoline price; or (2) simultaneous estimation of the two equations. The former approach was taken by Banaszak et al. (1999) for Taiwan and S. Korea, and the latter by Chandrasiri (1999) for Sri Lanka, who applied the SURE method[15] to the simultaneous equations:

$$Q_p = k_1 Y^{\alpha_p} P_p^{\beta_p} P_d^{\varphi} \qquad (6.1)$$

$$Q_d = k_2 Y^{\alpha_d} P_d^{\beta_d} P_p^{\phi}$$

where Q_p = gasoline demand (sales/#vehicles)
Q_d = diesel demand (sales/#vehicles)
k_1, k_2 = constants
Y = per capita GDP (at constant prices)
α_p = income elasticity, petrol
α_d = income elasticity, diesel demand
β_p = own-price elasticity for petrol
β_d = own-price elasticity for diesel
γ = cross-price elasticity of diesel demand to the gasoline price
φ = cross-price elasticity of gasoline demand to the diesel price

The results for the time series 1960–97 are given in Table 6.3. These suggest that substantial increases in price will be necessary to discourage consumer demand for auto fuel: a 10 per cent increase in gasoline price may reduce the demand for gasoline by less than 1 per cent (0.78 per cent) and increase the demand for diesel by 1.05 per cent. Similarly, a 10 per cent increase in diesel price may reduce the demand for diesel by 1.36 per cent and increase the demand for gasoline by 2.5 per cent. Indeed, these values are significantly below most other estimates for Sri Lanka.

Table 6.3 Elasticity estimates

Fuel	k	Income elasticities	Own-price elasticities	Cross-price elasticities	Corrected R^2
Gasoline	+1.76	$\alpha_p = 0.296$	$\beta_p = -0.078$	$\varphi = +0.250$	0.67
	(9.71)	(6.78)	(1.49)	(5.23)	
Diesel	+1.83	$\alpha_d = 0.129$	$\beta_d = -0.136$	$\gamma = +0.105$	0.88
	(8.80)	(2.58)		(1.75)	

Note: Figures in parentheses are t-statistics

Source: Chandrasiri (1999), Table 8.

However, the consensus view of the international literature favours the so-called "lagged endogenous model," which, say for the case of gasoline demand, has the general functional form[16]

$$Q(g)_t = kQ(g)_{t-1}^{\lambda} Y_t^{\alpha} P(g)_t^{\beta} P(d)_t^{\mu} V_t^{\gamma} \qquad (6.2)$$

where $Q(g)t$ = gasoline demand (sales/#vehicles) at time t
$P(g)t$ = gasoline price at time t
$P(d)t$ = diesel price at time t
k = constant
Y = per capita GDP (at constant prices)
V = vehicle mix variable (for example, fraction of motorcycles)
λ = lag coefficient
α = income elasticity
β = short-run own-price elasticity
μ = cross-price elasticity for diesel

from which the long-run price elasticity β^* computes to:

$$\beta^* = \frac{\beta}{(1-\lambda)} \tag{6.3}$$

For the gasoline equation, the relevant vehicle-mix variable is the fraction of motorcycles, while for the diesel equation, the fraction of light diesel vehicles is the relevant variable.[17] The results are shown in Table 6.4, where we compare the OLS and SURE estimates for the time series 1968–2000.

Table 6.4 Comparison of SURE and OLS estimates

Variable	Petrol		Diesel	
	OLS	SURE	OLS	SURE
CNST	0.412	0.365	1.94	2.08
	(0.226)	*(0.207)*	*(2.00)*	*(2.21)*
Petrol price	**-0.151**	**-0.155**	0.045	.039
	(1.82)	*(1.90)*	*(0.833)*	*(0.740)*
Diesel price	0.091	0.089	-0.070	-0.630
	(1.02)	*(1.01)*	*(1.18)*	*(1.06)*
Tax rates	-0.059	-0.054	**-0.144**	**-0.156**
	(0.317)	*(0.300)*	*(1.96)*	*(2.17)*
Lag effect	**0.704**	**0.693**	**0.765**	**0.758**
	(6.38)	*(6.35)*	*(7.46)*	*(7.62)*
Fraction of motorcycles	**-0.100**	**-0.099**		
	(2.18)	*(2.18)*		
Fraction of heavy duty vehicles			**-0.151**	**-0.168**
			(1.37)	*(1.54)*
R^2	0.962	0.962	0.979	0.979

Note: **Boldface** = statistically significant, t-value in parentheses

The lag variables have the correct sign and magnitude (for the hypothesized distributed lag model), and the results of SURE and OLS are surprisingly close. The vehicle fleet variables (fraction of motorcycles in gasoline, and fraction of heavy vehicles in diesel) are significant and also have the correct sign. However, we note that the diesel own-price elasticity is not significant, which is hardly surprising given the impact of vehicle taxes that may as much as double the cif price at time of purchase.

Of course, this finding is not very helpful in predicting the impact of increases in the diesel fuel price (see Chapter 7). However, as is evident from the diesel fuel price over the past two decades, with very little variation in real price (except a gradual downward drift), it is hardly surprising that an econometric estimate would have difficulty in identifying a clear price response.

The wide variation in the elasticity values reported in the literature is a reflection of the shortcomings of the data series used in these studies. Whatever values are used need to be treated with great care: and policy conclusions that depend upon them need to be subjected to careful sensitivity analysis. Thus we may summarize the elasticity values as shown in Table 6.5.

Table 6.5 Short- and long-run price elasticities

		Gasoline	Diesel
Own-price	Short-run	**-0.16**	-0.07
	Long-run	**-0.51**	**-0.29**
Cross-price		0.09	0.05

Note: **Boldface**= statistically significant

A different approach, though one that also requires a more detailed database, is to derive vehicle–km elasticities. These measure the change in vehicle–km travelled in response to changes in total (private) vehicle operating costs (including fuel, lubricants, and maintenance), and which will be different for different vehicle classes.[18] The results are shown in Table 6.6. These vehicle–km elasticities are used in the analysis of congestion tolls presented below.

Table 6.6 Vehicle-km elasticities

Vehicle Category	Elasticity
Cars (diesel/petrol)	-0.349
Light truck (diesel)	-0.230
Medium truck (diesel)	-0.443
Heavy truck (diesel)	-0.260
Medium buses (diesel)	-0.150
Large buses (diesel)	-0.130
Motorcycles (petrol)	-0.462

Source: Estimates by D. Jayaweera

Finally, we note the price elasticity of public transport fares. Previous estimates (Table 6.7) show low own-price and negative cross-price elasticities, suggesting that the different modes provide complementary rather than substitute services (Dheerasinghe and Jayaweera, 1997). The most significant bus fare increases (in constant terms) occurred in 1974, 1980, and 1987, with relatively little shift to rail services.

Table 6.7 Estimated price elasticities

Public bus	-0.08
Private bus	-0.07
Rail	-0.04

Source: Dheerasinghe and Jayaweera (1997)

Figure 6.6 shows average rail and bus fares per km during the 1990s, together with total passenger-km: monthly data is available only for 2001. Rail usage has increased significantly following the bus fare increases of 2000, climbing from 7 per cent of person-km to 9 per cent.

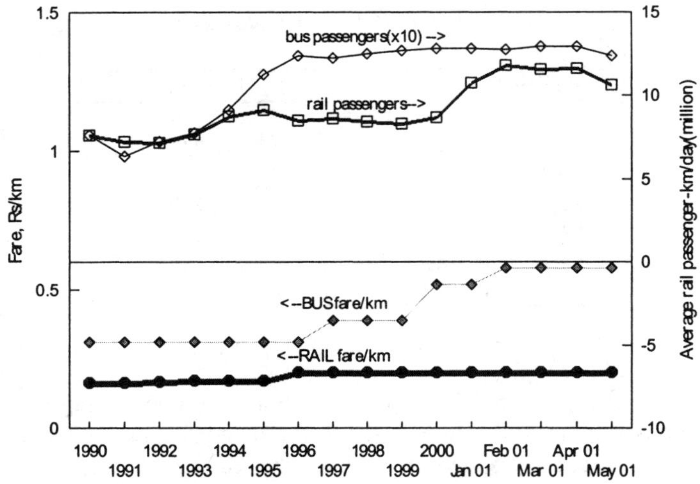

Figure 6.6 Bus and rail fares

6.2 THEORY

Figure 6.7 illustrates the fundamental relationship between speed, cost and traffic flow. In Figure 6.7A, starting at low flows (V_1) and an empty road, speed is S_1. As additional vehicles join the flow, speeds gradually decline, reaching S_{max} at the flow V_{max}. Beyond this point, additional vehicles entering the road result in a reduction in flow, reaching near zero flow and near zero speed (S_o) as total gridlock approaches.

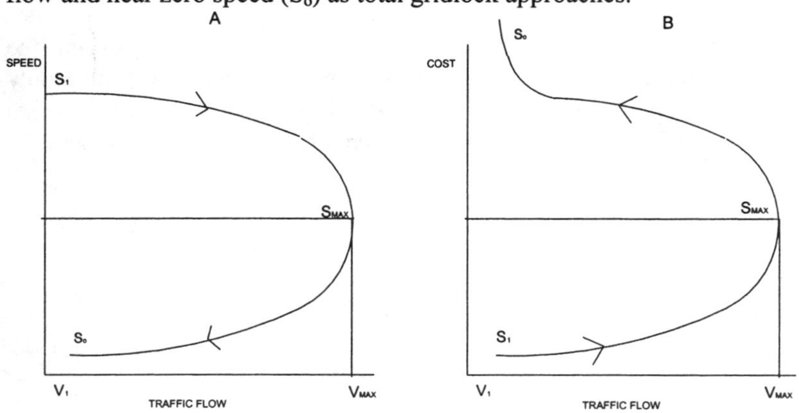

Figure 6.7 Speed and cost v. traffic

Figure 6.7B shows the corresponding average private cost (APC) curve faced by the individual motorist.[19] As speeds decline, private cost to the motorist increases (because as speeds fall, time per trip, and the value of time per trip increase; and because as speeds decline, fuel efficiency decreases). As more vehicles enter the road beyond V_{max}, costs continue to increase (tending to infinity as total gridlock approaches). In the subsequent discussion of theory, in the interests of clarity we do not draw the backward bending part of the cost curve – optimal solutions will never lie in this domain, for they will always lie on the lower part of the cost curve.

Traffic congestion occurs because the marginal user considers only his or her own private costs when making a decision to undertake a particular trip at a particular time, ignoring any costs imposed on others. Figure 6.8 illustrates the theory. Up to some given traffic volume (V_o), equal to the capacity of the road, the private cost is (roughly) constant (P_o). Below this value, an additional vehicle does not reduce the speed of other vehicles already on the road.

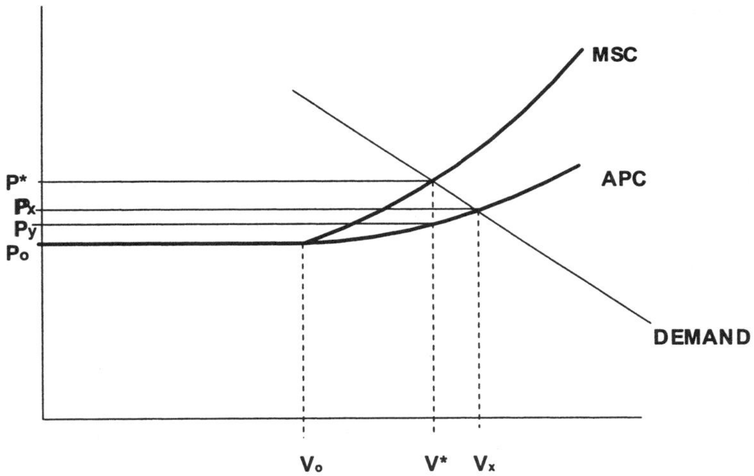

Figure 6.8 Optimal pricing and congestion

But with further traffic increases, the volume to capacity ratio exceeds one, speeds begin to decrease, and therefore the individual user's private costs increase – because to cover a given distance requires more time (which has a cost to the individual), and because fuel efficiency is lower as speeds decrease (which increases the individual's fuel costs). Therefore, beyond V_o, the private cost (APC) curve slopes upward, intersecting the demand curve at V_x, P_x.

Beyond V_o, however, the marginal user imposes these same externalities on the other road users as well, for the decrease in speed is experienced by all. The marginal social cost (MSC) curve therefore lies higher than the private cost (APC). Thus the societal optimal traffic volume is at V*, P* – where the demand curve intersects the MSC (rather than APC). Welfare theory argues for a "Pigouvian" tax[20] on the use of congested roads in the amount $\tau = P^* - P_Y$, at which point the private users' total cost (private cost plus tax) exactly equals the social cost (and which also equals the marginal social benefit represented by the demand curve).

Traffic demand during off-peak hours is typically very much smaller, implying that the off-peak demand curve lies far to the left and that the optimal toll would be correspondingly smaller, if not zero, as indicated in Figure 6.9. In fact, the present situation is that before 7 am, and after 9 pm, the actual traffic is well below capacity, which means that the optimal toll for this period is zero.

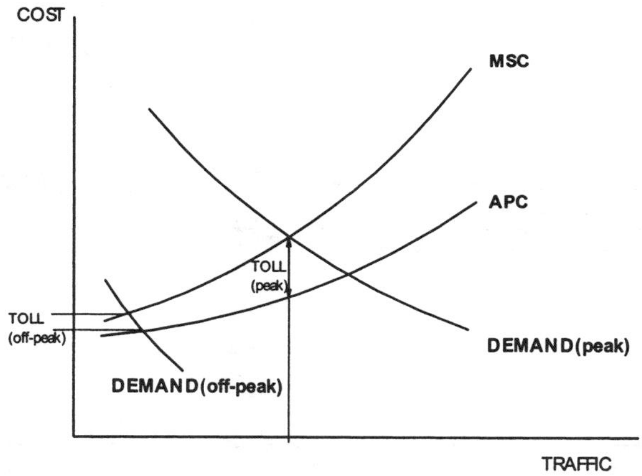

Figure 6.9 Optimal tolls, peak v. off-peak

6.3 THE VALUE OF TIME: EMPIRICAL ESTIMATES

Time savings account for the largest proportion of economic benefits of reducing traffic congestion. Both absolute as well as relative values are of interest, since different road users – drivers, bus passengers, car passengers – have significantly different incomes, and would likely value their time at different levels.[21]

Table 6.8 shows average monthly incomes of drivers and passengers, which may be used as a rough guide for relative values. In subsequent sections we use the value of car passenger time as the exogenous variable, with that of other users linked by the proportions implied by the values in Table 6.9. The car passenger value of time is escalated at the assumed real GDP growth rate (which we take as 5 per cent per year in the reference case).[22]

Table 6.8 Average monthly income of drivers and passengers

	Rs/month
Car passenger	21,000
Van passenger	16,500
Bus passenger	3,800
Car driver	4,700
Van driver	5,040
Bus driver	10,080
Truck driver	11,280

Source: University of Moratuwa, Origin–Destination Survey, May 1998

Table 6.9 Trip purpose, morning peak

	%
Work (commuting)	43
Education	21
Business	16
Recreation	5
Social	9
Other	6

Source: University of Moratuwa, Origin–Destination Survey, May 1998

If we escalate the passenger salary by 5 per cent, the 2001 value is 24,310 Rs/month, equal to 138 Rs/hour.[23] This definition of the value of travel time appears to have been used by the Colombo Urban Transport study (Halcrow-Fox Associates, 1996), for their estimate (at 1998 prices) is 111.18 Rs/hour for car passengers.[24]

Several studies suggest that the value of time is related to trip purpose (Table 6.10), for which some data are available in Sri Lanka: the value of time for business (and commuting) is arguably greater than leisure time.[25] Kumarage summarizes the value of time savings as shown in Table 6.10.

Table 6.10 Value of Travel Time Savings (in 1999 Rs)

User group	Urban	Rural	Intercity	All sectors (a)
Car	100.06	78.62	135.81	106.50
Van	51.15	37.62	51.15	48.44
Motorcycle	19.05	27.00	14.29	19.22
Public transport	10.83	12.41	12.41	11.62
Non-motorized modes	6.78	8.62	0.00	7.39
All motorised modes (c)	24.61	23.01(b)	28.81	25.55

Notes:
(a) Assumed as 50 per cent urban, 20 per cent rural and 30 per cent intercity travel
(b) If non-motorised travel is considered, for example in rural transport, this should change. If 50 per cent of travel is assumed to be by non-motorised means, then the average value of travel time savings would be Rs 15.31 per hour.
(c) Assumed as 10 per cent share for cars, vans and motorcycles respectively and 70 per cent for public transport.

However, Jayaweera's estimates of the values of time are substantially lower, as shown in Table 6.11. The average value of time (for car passengers) is only Rs26.5/hour (Rs30/hr at 2000 price levels).[26] This is significantly below the Rs111/hour value used by the Colombo Urban Transport Study, and below the value of Rs66/hour used by the CKE-EIS.[27]

Table 6.11 Value of Time

	Value of time(a) Rs/hour	Ratio to value of time for car passenger	Ratio based only on salary
Car passenger	26.5	1.00	1.00
Van passenger	21.4	0.81	0.79
Bus passenger	8.8	0.33	0.18
Car driver	16.8	0.63	0.22
Van driver	13.0	0.49	0.24
Bus driver	16.4	0.62	0.48
Truck driver	19.3	0.73	0.54

Note: (a) The source document expresses the value of time per one-km length trip at a speed of 42 km; therefore we multiply those figures by 42 to obtain values per hour.

Source: Jayaweera (1999), Tables 5.8 and 5.2

The average value of time for goods vehicles was provided by the TSPC as 186Rs/hour for industrial inputs, 154Rs/hour for intermediate goods, and Rs196/hour for finished goods. To take into account some proportion of trips made empty, we have used an average of Rs 150/hr (to which we add driver's time as noted above).

6.4 IMPACT OF A CONGESTION TOLL

Table 6.12 shows the graduated toll proposal of Jayaweera and the flat rate proposal of the Colombo Municipal Council.[28] Jayaweera's proposal for a graduated toll is based on his theoretical analysis that takes into account the differences in externality costs imposed by different vehicle classes – and is used in this volume as the reference case. Clearly there is some uncertainty about whether the response to a toll will conform to the historically estimated responses to changes in VOC.[29]

Table 6.12 Proposed tolls (daily equivalent)

	Jayaweera	Colombo Council
Car	25	8
Light Truck	20	8
Medium Truck	20	8
Heavy Truck	40	8
Medium Bus	0	
Light Bus	0	
Motorcycles	10	8

The model used in this study to estimate changes in vehicle speeds as a function of changes in traffic volumes (which are in turn influenced by the value of the toll and the assumed price elasticity) is described in detail in Meier and Munasinghe (2003b). Figure 6.10 shows the predicted changes in road speeds in response to a toll, based on the elasticities of Table 6.6.[30]

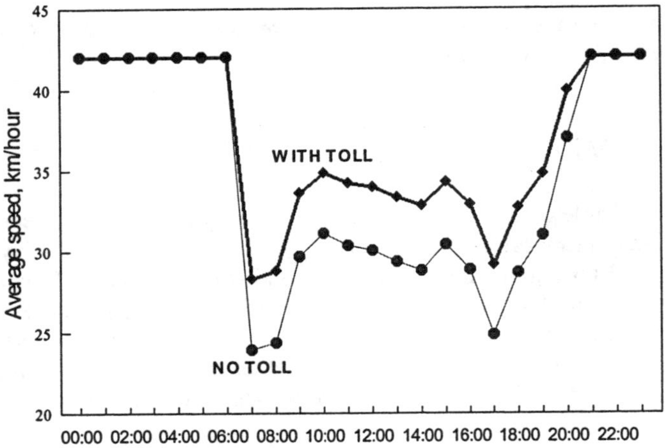

Figure 6.10 Impact of congestion toll on speed (A3, outbound)

The corresponding pattern of avoided trips is shown in Figure 6.11. However, these are offset by a small number of additional bus trips (not shown in the figure) induced by the toll – a ratio of roughly one additional bus trip for every 40 avoided motorcycle or passenger car trips[31] – but these have little effect on overall vehicle numbers.

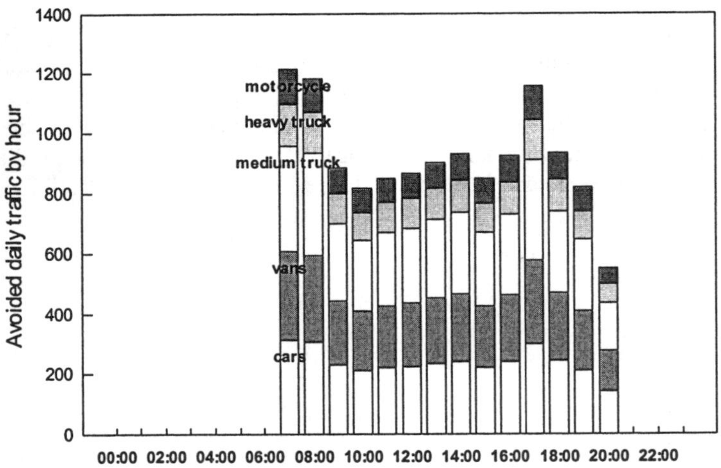

Figure 6.11: Avoided traffic (basecase tolls, A3)

6.5 ECONOMIC ANALYSIS

The calculation of the main economic and financial flows is straightforward, but care must be taken to account for welfare losses that depend upon behavioural assumptions in response to the toll. In the case of passenger trips (cars, vans and motorcycles), there are four potential responses to a congestion charge:

1. Make the trip at the usual hour, and pay the toll;
2. Forgo the trip entirely;
3. Take public transport (that is, shift from car, van or MC to buses);[32]
4. Avoid the toll by entering the city before 7 am, remain in the city, and leave the city after the toll period ends at 8 pm. However, since the toll will not be collected at booths, but instead require the display of stickers during the peak hours, the requirement to be parked (and not be observed on the road) is onerous, and it may safely be assumed that the magnitude of this response would be quite small (and is ignored in the model).

Figure 6.12 shows the loss of consumer surplus implied by the reduction in the number of trips in response to the toll. The total loss of surplus is the area Y + X, namely

$$X = \frac{\tau[N_0 - N_{toll}]}{2}$$

$$Y = \tau N_{toll}$$

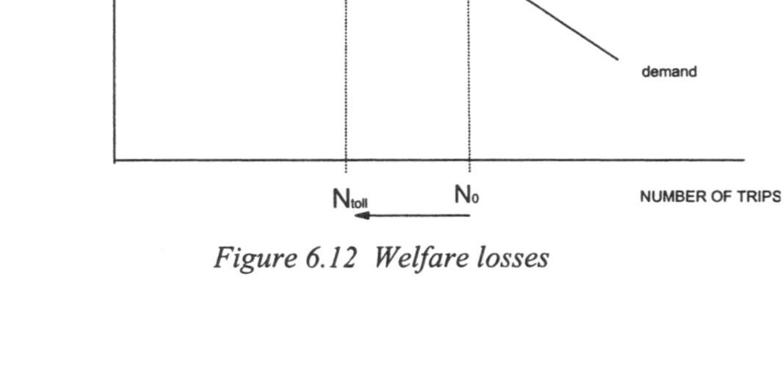

Figure 6.12 Welfare losses

where τ is the daily toll, and N_o and N_{toll} are the number of trips before and after the toll, respectively. Thus *Y* is the toll revenue collected by the government, and *X* is the deadweight loss.

This can be taken only as a rough approximation of the various effects, which are quite complicated in the case of those who shift to public transport. Since the original decision to commute by car or motorcycle is motivated by a combination of the desire to avoid the longer trip times associated with public transport, to enhance personal comfort and convenience, and to display status, those who shift to public transport will experience some loss of utility (which may or may not be offset by the lower cost of public transport).[33]

Table 6.13 shows the physical environmental impacts as a function of key assumptions: the level of the toll and the assumed elasticity to the toll. It is evident that if the road user response to tolls is the same as the historical VOC elasticity, then the reductions in traffic in response to a Rs 25 daily toll will be modest, no more than 2.2 per cent in total, with a 3.9 per cent reduction in car traffic, and a 5.6 per cent reduction in motorcycle traffic. The same is true of the reduction of air emissions, which is small. To obtain, say, a 20 per cent reduction in car traffic would require a Rs 75 toll, and a much greater response to the toll than the estimated VOC elasticities suggest.

Table 6.13 Physical impacts of the congestion charge

		Toll = Rs 75		Toll = Rs 25	
		Low elasticity	High elasticity	Low elasticity	High Elasticity
Cars	(%)	-10.1	-19.1	-3.9	-7.7
Light truck	(%)	-7.9	-15.0	-3.1	-6.1
Medium truck	(%)	-9.7	-18.3	-3.7	-7.2
Heavy truck	(%)	-9.7	-18.3	-3.9	-7.6
Medium buses	(%)	0.3	0.5	0.1	0.2
Large buses	(%)	0.6	1.0	0.3	0.5
Motorcycle	(%)	-14.0	-25.7	-5.6	-10.8
Total	(%)	-5.7	-10.7	-2.2	-4.3
Environment (emission reductions)					
GHGs (lifetime)	1000 tons	17.0	33.0	6.4	12.8
Year 2010 reductions					
PM-10	(tons)	1	2	0	1
SO_2	(tons)	7	13	3	5
NOx	(tons)	39	75	15	29

Notes:
Low elasticity = values per Table 6.6.
High elasticity = 2 x VOC values.

Table 6.14 shows the distribution of costs and benefits for the reference case, under the toll structure proposed by Jayaweera. It is evident that the big winners are (low income) bus passengers – because of the large number of passengers in each bus, increases in bus speeds have disproportionately large impacts on aggregate time savings, and therefore on economic benefits as a whole.

Table 6.14 Impact of the congestion charge, year 2010 (passenger car toll = Rs 25)

	Road users									
	Low income (Bus)	Med income (MC)	High income (Car)	Freight (Lorry)	Govt	Others	Total	Environment Local	Environment Global	Total
	[1]	[2]	[3]	[4]	[5]	[6]	[7]	[8]	[9]	[10]
Time savings	578	21	489	354			1,442			1,442
Tolls	0	-99	-1,280	-593	1,972		0			0
Gasoline fuel tax		0	8		-8		0			0
Diesel fuel tax	5		1	2	-9		0			0
Diesel savings	87			47			134			134
Petrol savings		0	19				20			20
Deadweight losses	0	-7	-56	-27			-90			-90
Admin. costs					-30		-30			-30
Commissions					-197	197	-197			-197
Environment										
PM10							0	2		2
SO_2							0	2		2
NOx							0	3		3
GHG							0		1	1
Total	671	-85	-819	-217	1,728	197	1,278	7	1	1,286

If deadweight losses are excluded, the effect on car passengers is about even: that is, their own time and fuel savings are about the same as the toll payment; but if deadweight losses are added – that is, the loss of surplus for those car users who shift to a less convenient transport mode – then the net impact on present car owners is negative. In the case of trucks, even with the estimated deadweight losses included, the net impact is positive, so there is no case that the imposition of tolls would increase commodity prices.[34] Figure 6.13 shows the corresponding results in net present value terms over 20 years (at a 10 per cent discount rate).

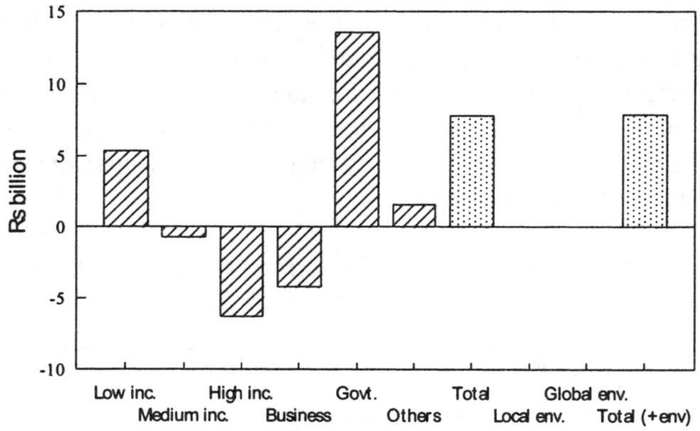

Figure 6.13 Congestion toll winners and losers (NPV over 20 years)

It is evident that road pricing in the form of a congestion toll has a significant positive economic impact. The inclusion of environmental benefits only raises benefits by less than 1 per cent; however, as a win–win option these environmental benefits are achieved at no incremental cost.

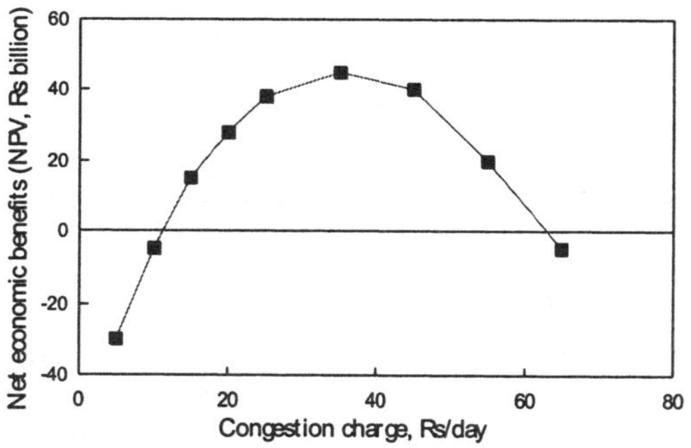

Figure 6.14 Net economic benefits v. congestion charge

Figure 6.14 shows a sensitivity analysis of net economic benefits to the level of the congestion charge (for 2001).[35] The optimum corresponds to a passenger car toll of about Rs 35/day – which is somewhat higher than the theoretical value of Rs 25 determined by Jayaweera on the basis of a microeconomic analysis. The reason is that buses are exempted in the policy proposal, so the toll for remaining vehicles has to be increased somewhat to have the same overall impact on congestion. However, the results of this aggregate (macro) model clearly confirm the theory.

6.6 OTHER OPTIONS FOR REDUCING CONGESTION

Meier and Munasinghe (2003b) examine a range of other options for reducing traffic congestion in the Colombo area: the construction of an expressway from the centre of Colombo to the Katanuyake airport; railway electrification; the reintroduction of school buses; and a dry container port, designed to reduce congestion not just of the main northbound artery, but also in Colombo Port itself. Each was analysed using the same methodology as described above for road pricing.

The Katunayake Airport Expressway

A four-lane, limited access expressway to link downtown Colombo with the Katunayake airport – a total distance of some 25 km – has been discussed for almost two decades.[36] The present A3 road to the airport is the most heavily travelled in Sri Lanka, with average daily traffic (ADT) in the sections closest to Colombo of 200,000: this section of the A3 accounts for some 1 per cent of all vehicle-km in the country.

The original feasibility and environmental impact studies provoked many objections,[37] and the project has been repeatedly delayed. Consequent to delays, construction costs have also increased, with the most recent estimate now standing at Rs 13.1 billion, up from the Rs 5 billion estimate of 1995. These cost increases have of course raised doubts about the economic analysis of the original proposal. However, construction finally began in 2000.[38]

While opponents of the project have pointed to the environmental impact associated with construction, the air pollution impact of the present situation is becoming more serious as traffic congestion and travel times on the present A3 road have grown. Since emissions per kilometre of both local air pollutants and carbon are inversely proportional to speed, in the absence of the expressway emissions will increase due to yet further expected decreases in vehicle speeds.

Although the decision to proceed with construction of the expressway has now been taken by the Government, there are a number of other proposals for projects of a similar type, notably the proposed Southern Highway (linking Colombo to Matara on the South Coast), and the Outer Circular Highway (Colombo orbital road, which would also connect to the Airport expressway). Insights derived from the analysis of the expressway project may be taken to apply to these proposals as well.[39]

Theory

The economic benefits of a new road can be calculated as the change in consumer surplus enjoyed by its users.[40] Consider the demand curve D_{T1} depicted in Figure 6.15. Before the project, users of the A3 experience costs of P_{A3} at the traffic volume Q_{A3} (given by the intersection of the demand curve and the marginal cost curve, MC_{A3}).

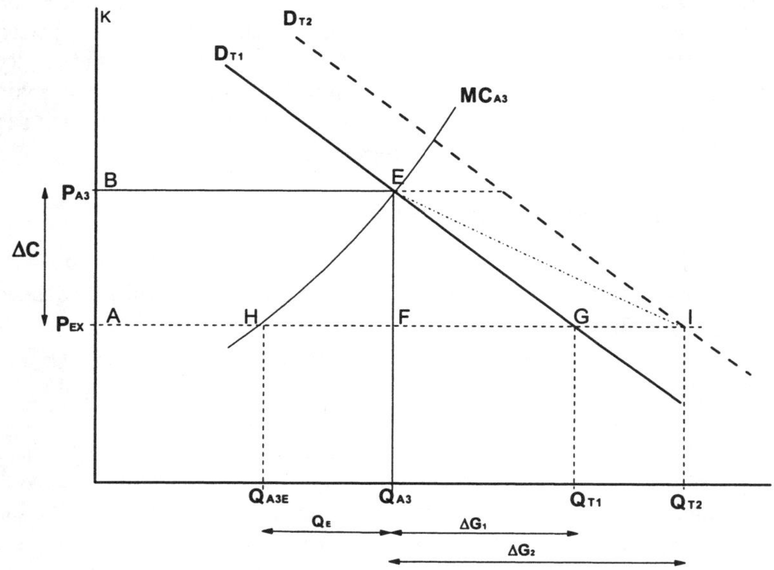

Figure 6.15 Estimating consumer surplus changes

Once the project is built, some vehicles (Q_E) will elect to use the expressway, leaving Q_{A3E} on the original road. Road users will experience lower costs, P_{EX}, due to fuel and time savings $\Delta C = P_{A3} - P_{EX}$. However, at the new lower cost P_{EX}, the total travel demand increases from Q_{A3} to Q_{T1}, the difference ΔG_1 being the induced traffic (whose quantum is a function of the demand elasticity, that is, of the slope of the line D_{T1}).

The benefit is the change in consumer surplus, which increases from the original area KBE to KBAHFGE.[41] Thus:

$$\Delta B_1 = KBAHFGE\text{-}KBE = ABEF + EFG$$
$$= \Delta C\, Q_{A3} + 0.5\, \Delta C\, \Delta G_1 \qquad (6.4)$$

There is also likely to be an outward shift in the demand curve, shown in Figure 6.15 as D_{T2}. This intersects price P_{EX} at I, from which follows an additional increase in traffic to Q_{T2}. The result is a slightly larger increase in consumer surplus, namely from KBE to KBAHFGIE, hence

$$\Delta B_2 = KBAHFGIE - KBE = ABEF + EFGI \qquad (6.5)$$
$$= \Delta C\, Q_{A3} + 0.5\, \Delta C\, \Delta G_2$$

However, at the higher level of traffic within the corridor, there will inevitably arise increased congestion outside the corridor, particularly on the access roads to the expressway interchanges. This externality, say ΔC^*, would reduce the level of benefits. In this study we make the assumption that this externality exactly offsets shifts in the demand curve induced by the expressway,[42] so that benefits are estimated as the amount ΔB_1 (in Equation 6.4).

As one might expect for a project whose planning has extended for over a decade, many traffic forecasts have been made, and under all kinds of assumptions. The most recent forecast appears to be that prepared by the University of Moratuwa (2000) for the Board of Investment (BoI) which examined nine scenarios based upon different assumptions about GDP growth, automobile ownership, level of toll (among others). Traffic forecasts were developed with use of the TransPlan model (Kumarage et al., 2000).[43] A toll price elasticity of -0.2 is used in the analysis, based on this University of Moratuwa report.[44]

Not surprisingly, the resulting traffic estimates span a very large range. For example, the 2025 estimates of daily vehicle-km on the expressway range from 325,000 to 747,000, and on the A3 from 836,500 to 1,708,000; the corresponding toll revenue varies from Rs 1.7 to 4.1 million/day.[45]

The earlier expressway traffic forecasts appear to be based on a flat toll of Rs 50. However, the University of Moratuwa study is based upon the toll structure shown in Table 6.15.

Table 6.15 Tolls proposed by BoI

Vehicle type	Rs
Car	65
Van	95
Bus	140
Large bus	180
Delivery van	195
Truck	390
Container truck	650

Note: One-way toll (Rs) in 2000, Peliyagoda to Katunayake

These are high tolls, particularly for trucks. In the case of cars, Rs 65 is 50 per cent of the salary value of car passengers, and in line with potential time savings (of 30 minutes, or up to 90 minutes at rush hour) for a trip to the airport. However, the toll is significantly higher than the value of time for goods vehicles.

Table 6.16 shows a comparison of the various valuation estimates of environmental impacts for the CKE. For some categories of impacts (for example, water pollution, noise, reduction in road accidents and reduction in fishery productivity[46]), we have adopted the estimates of the CKE-EIS and Sivagnasothy and McCauley (2000), as indicated; for local air emissions we use our own estimates.

Table 6.16 Valuations of local environmental impacts NPV, Rs Million

	Sigvagnasothy and McCauley (1991 prices)	CKE-EIS (1995 prices)	Updated to 2000(a)
1. Conversion of agricultural land(b)	-42		-84
2. Noise and air pollution(c)	-41		
3. Noise pollution(d)		-29	-40
4. Air pollution		1,377	See text
5. Reduced biodiversity and area of lagoon		-83	-118
6. Water pollution and reduction in fishery productivity	-8		-16
7. Reduction in road accidents(e)		112	159
8. Conversion of non-agricultural lands	-660		-1,331
9. Loss of structures	-598		-1,206
10. Increased flooding		-70	-99

Notes: Negative numbers = costs; positive numbers = benefits

(a) GDP deflator from Central Bank Annual report

(b) Opportunity costs of lost agricultural production

(c) This estimate is based on a simple hedonic pricing method, in which noise and air pollution impacts have been estimated on the basis of estimated changes in property values

(d) Assuming a 10 per cent loss of property value of the 416 properties expected to be affected by increased noise of the expressway

(e) The CKE-EIS reports the figures on an annual basis. These are converted to NPVs over 25 years at 12 per cent discount rate

Items 8 and 9 in Table 6.16, concerning conversion of non-agricultural lands and structures which were inadequately treated by the earlier EIA, appear to have been included in the new estimate of capital cost (which reflects more generous compensation packages than originally contemplated). We have already noted the error in the CKE-EIS valuation of local air pollution benefits, which is obviously overstated (and therefore replaced by our own calculations).

The distribution of costs and benefits are shown in Table 6.17 and Figure 6.16. Bus passengers are assumed to benefit only from time savings (since fuel savings would accrue to bus owners): bus tolls would be paid by owners, and are assumed to be recovered from fuel savings rather than charged to passengers in the form of higher fares.[47]

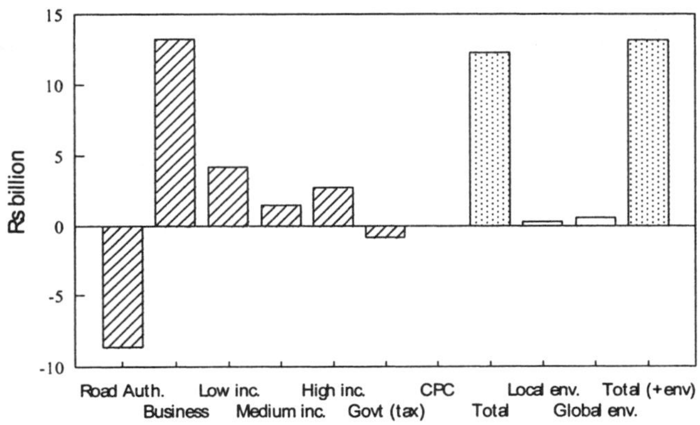

Figure 6.16 Distribution of costs and benefits by stakeholder group

Even though motorcyclists would be banned from the expressway (and therefore incur no tolls), they benefit from higher speeds on the A3 (deriving both time savings and fuel savings). It is assumed that any time savings of bus, van and truck drivers accrue to the owners of these vehicles. It is evident from Table 6.17 that the air emission benefits outweigh the non-air costs (as computed by others), and that, in this case, consideration of the environmental benefits simply makes a good project better.

As expected, freight is the single largest beneficiary. Perhaps less expected is the result that low-income road users, as a group, derive greater benefit than car passengers. The reduction in road accidents is shown under low income, for the majority of fatalities and serious injuries are suffered by pedestrians, who – at least in the case of the A3 – may be assumed to fall into lower rather than higher income brackets. It may be seen that the toll income (even at a Rs 50 flat toll) would more than cover O&M costs of the Road Development Authority (RDA), but still leave a significant funding gap for debt service.

Table 6.17 Distribution of costs and benefits (NPVs in Rs Million at 2000 price levels)

		Road users									
			Low	Med	High				Environment		
	RDA	Freight (Lorry)	income (Bus)	income (MC)	income (Car)	Govt	CPC	Total	Local	Global	Total
	[1]	[2]	[3]	[4]	[5]	[6]	[7]	[8]	[9]	[10]	[11]
Tolls	2,922	-2,922			-841			-841			-841
Construction	-10,372							-10,372			-10,372
O&M costs	-1,140							-1,140			-1,140
Time savings		14,505	4,065	1,371	3,051			22,992			22,992
Diesel fuel tax						-419	419	0			0
Petrol fuel tax						-359	359	0			0
Diesel savings		1,681					-1,681	0			0
Petrol savings		-0		88	536		-624	0			0
Fuel imports							1,472	1,472			1,472
Port distrib. margins							55	55			55
Road accidents			115					115			115
Environment											
Non-air impacts									-357		-357
Air emissions											
PM-10								0	147		147
SOx								0	253		253
NOx								0	290		290
Carbon								0		574	574
Total	-8,589	13,264	4,180	1,460	2,746	-778	0	12,281	333	574	13,188

Railway Electrification

The Sri Lanka Railway has been studying the possibility of electrifying the suburban railways in Colombo for several decades. Objectives have changed over time ranging from reduction of fuel imports (particularly during the fuel crisis of the 1970s, when hydro power generation was in surplus during non-peak hours) to more recent concerns of attracting passengers from road to rail arising from the need to alleviate road traffic congestion. A decision is now being considered by the Government, which is looking for private sector investment for railway electrification.[48]

The section of railway line proposed for electrification between Colombo Fort and Veyangoda is part of the main line. The benefits of railway electrification can be enumerated as follows:

- Travel time savings to former diesel multiple unit (DMU) passengers: the electric multiple units (EMU), by virtue of the higher travel speeds and greater operational frequency, will save time for the former DMU passengers. This saving in time is calculated by deducting the passenger hours by EMU travel from the corresponding value for DMU travel. The total time saving is then multiplied by the value of time taken as Rs 10.83 per hour.[49] This value is assumed to increase in real terms with the increase in per capita income assumed for each scenario.

- Travel time savings to former bus passengers: these are calculated by taking the difference in travel times for those bus trips that have now been diverted to the EMUs. The same value of travel time has been applied for these passengers as well.
- Vehicle operating cost savings from diversion of former bus passengers to EMUs:[50] the reduced demand for bus passengers resulting from the transfer of former bus passengers to rail would then lead to fewer buses having to operate on the congested road network. The operational economic cost of these buses is taken as Rs 18.80 per km. A further assumption is made that each bus carries on average 50 passengers.[51]
- Reduction of air emission damages from reduced use of buses on congested roads (see Table 6.18).
- Reduction of air emissions due to electrification: under the assumption that the incremental electricity required for EMU operation is based on oil or coal, and that roughly the same amount of fossil fuel would need to be used for that electricity generation as is presently consumed by the DMUs, even if there is no change in the level of emissions, there is a change in the *location* of emissions.[52] DMU emissions occur at trackside which for much of the suburban route passes through densely populated areas, whereas power generation, particularly if it occurs at a remote coal plant, results in emissions released from a tall stack – for which the health damage estimate is an order of magnitude less than for ground level emissions in populated areas (Lvovsky et al., 2000, Table 6.3). Typical health damages for high stack and ground level emissions are in the ratio of 3,114 : 42 for PM-10, 487 : 6 for SO_2, and 123 : 2 for NOx. We apply these ratios to the emissions from former DMU operations (that is, to the 4.6 million litres of fuel estimated for the present DMU operations):[53] these are shown in Table 6.18 under DMU damages.
- Savings from accident costs from reduced bus travel: road transport accident costs have been estimated at Rs 0.08 per passenger-km (Kumarage, 1999). Although accident costs for rail transport are not available for Sri Lanka, an approximate computation using the same values for deaths and casualties as used in the road estimates gives a value of Rs 0.04 per passenger km. This is the cost for present railway operations. An increased level of safety will not be claimed for EMU operations, as costs for such have not been included in the project.

Table 6.18 Physical environmental impacts of railway electrification

		Road	Power generation		
		traffic	DMU (diesel)	EMU (coal)	Net
PM-10	tonnes	1.0	3.5	-0.4	3.1
SO_2	tonnes	11.4	385	-892	-508
Nox	tonnes	48.6	156	-74	81
Carbon	tonnes	1045.8	3,365	-4,461	-1,095
Fuel consumption	tonnes		3,846	-7434	

Note: positive value = benefit; negative value = cost

Railway electrification is found to be uneconomic (Figure 6.17), with a net economic loss (as NPV) of some Rs 3 billion. While there are clear benefits to road congestion and reduced vehicle operating costs, these are insufficient to warrant the high capital costs of electrification.

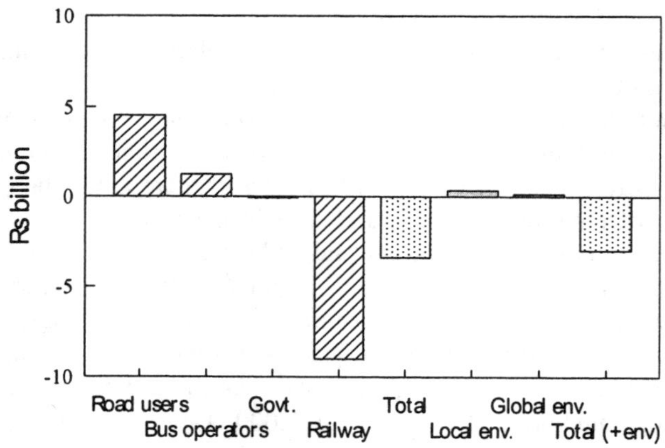

Figure 6.17 Winners and losers: railway electrification

Using a longer 30-year time horizon (that may be appropriate for electrification projects), the switching value of the discount rate is 4 per cent (with environmental benefits), or 3.5 per cent (without environmental benefits). These are much below the opportunity cost of capital (even if this were assessed at 7–9 per cent rather than the 10 per cent used for other options).

The environmental benefits, though positive, are extremely small by comparison, and are clearly insufficient to offset the economic loss before consideration of environmental damage costs. The environmental benefits from electrification per se, even when based on coal-fired generation, are higher than the air emission benefits of reduced road congestion – except for carbon emissions, which are higher from coal-fired generation.[54] Note that even though SO_2 emissions are greater from the coal power plant than from the DMU units, the health damage is lower because of the location of emissions.

This analysis confirms the general view of previous studies that the case for electrification is doubtful: our analysis shows that electrification of the northern commuter line is uneconomic. However, we also show that the environmental benefits are very small compared to the costs of electrification, even when we take into account the reduction in damage costs consequent to incremental EMU power generation at remote locations. Compared to other methods of reducing local air emissions, railway electrification is one of the most expensive options.

Dry Container Port

Colombo Port plays a key role in the economy of Sri Lanka. The handling of transhipment traffic is highly profitable, and the Port Authority has plans for further growth. However, there are significant harbour space constraints that prevent large-scale expansion, and therefore efficient operation of the port is of utmost importance. Colombo competes with several other ports in the region, each striving to become the regional hub for major container operators and consortia: most important, ports in south India endeavour to attract operators to use their ports directly without relays from others in the region. Nevertheless, Colombo – at the southernmost point of South Asia – has a unique geographical advantage, and consequently the Government is desirous of exploiting that advantage for transhipment operations.

Several recent studies have examined port operations, all of which have drawn attention to the length of time that domestic containers spend in the port, thereby occupying space that might otherwise be given to the more profitable transhipment traffic – which will become increasingly affected as the growth of the domestic economy requires increased domestic container traffic.[55] Even with improvements in productivity to reduce dwell time within the port, capacity is inadequate. To achieve further reductions in dwell time it has therefore been proposed that all domestic containers be moved to a port inland clearance depot (PICD), which would ensure that the present capacity of the port was sufficient to 2006.

The problems of congestion inside the port are aggravated by congestion in the surrounding City of Colombo. As domestic container traffic has grown, increasing numbers of articulated lorries move containers to their private inland container depots, many of which are located to the north of Colombo on the A3 radial road. This has significantly worsened congestion both on this key road, as well as on the local roads leading to the A3 (see Figure 6.18). Heavy lorries, which are almost all container vehicles, account for about 5 per cent of vehicles, but 8 per cent of road capacity (when measured in PCUs).

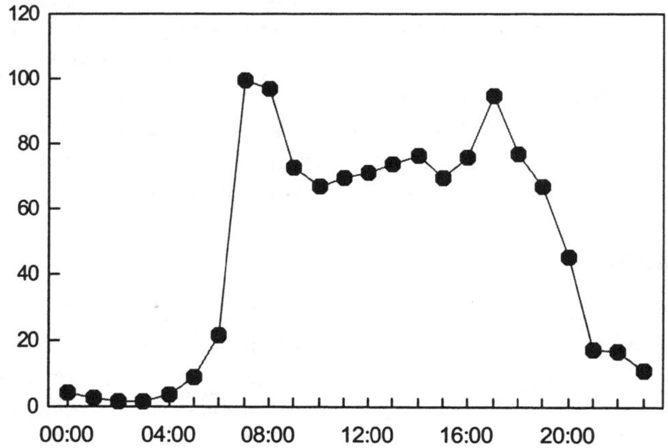

Figure 6.18 Heavy lorries, outbound A3 (2000): traffic counts by hour

Estimating the benefits on the operations of the Port of Colombo goes beyond the scope of this study – which will need to await the full feasibility study. Indeed, as noted, neither the pre-feasibility study nor the TPSC economic analysis attempted such a quantification. Nevertheless, based on interviews with private operators conducted during the pre-feasibility study, there is wide support among container operators for the idea, with preliminary indications that private sector involvement in financing the project would be likely. These benefits include:

- increased throughput of trans-shipment container traffic;
- deferral of the construction of a new port, or major new expansions to increase port capacity;
- cost savings to private operators attributable to reduction in dwell time in the port.

However, the model presented earlier for evaluating congestion pricing can be used to identify the benefits from reduced road congestion on the A3 (as container traffic is diverted to the railway), and the concomitant environmental benefits (reduction of diesel emissions of container trucks, and reduction of emissions from other road vehicles consequent on increases in average speeds once the container lorries are removed from the local roads and the A3). Removal of this traffic from the A3 would improve average speeds by about 3–4 km/hour during peak hours (Figure 6.19).

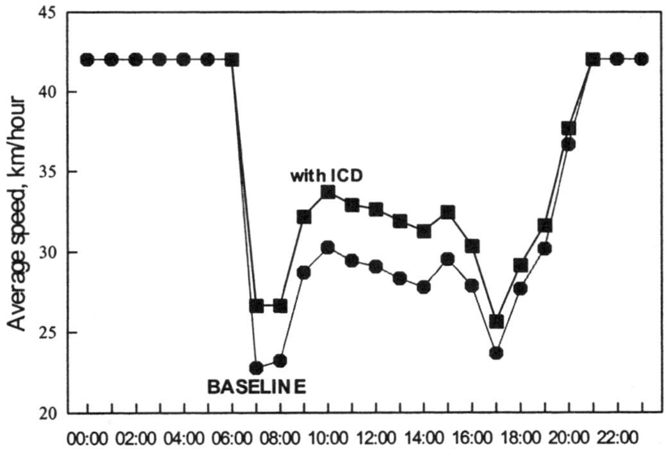

Figure 6.19 Impact of the PICD on speeds on the A3 (2001)

Table 6.19 shows the distribution of costs and benefits. The analysis considers both the A3 and Kandy roads (which are the two radial roads with the heaviest container lorry traffic): it is assumed that the PICD would reduce heavy lorry traffic on the A3 by 75 per cent, and on the Kandy Road by 50 per cent.[56]

Table 6.19 Costs and benefits of the PICD

	Road users								Environment		
	Low income (bus)	Med Income (MCs)	High income (cars)	Freight (lorry)	COs	Govt	SLR+ Port	Total	Local	Global	Total
Time savings	5,927	18	927	647	2,639			10,157			10,157
Tolls	0	0	0	0		0		0			0
Gasoline fuel tax		1	26			-27		0			0
Diesel fuel tax	44		4			-48		0			0
Diesel savings	765			273	63			1,101			1,101
Petrol savings		1	85					86			86
Deadweight losses	0	0	0	0				0			0
VOC savings					2,957			2,957			2,957
Fees to SLR/ICD					-6,579		6,579	0			0
Other efficiency savings					919			919			919
Capital cost							-1,712	-1,712			-1,712
O&M costs							-4,866	-4,866			-4,866
Environment											
PM-10								0	7		7
SO_2								0	11		11
NOx								0	13		13
GHG emissions								0		12	12
Total	*6,736*	*19*	*1,042*	*920*	*0*	*-75*	*0*	*8,642*	*31*	*12*	*8,685*

Note: CO = container traffic operators

We make the further (conservative) assumption that the residence time of rail shipment is 50 per cent of the corresponding value of road shipment.[57] It is assumed that the costs of the PICD incurred by the railways and the port are recovered in fees from the container operators, and 'other savings' have been set so that the assumed ERR of the project equals the opportunity cost of capital: these are assumed to be the minimum level of additional benefits to the container operators to make the project viable; and thus the net economic benefits shown are really equal to just the externality reductions – to road users and to the environment. Clearly, the former benefits are very large, the latter very small: and the principal beneficiaries are bus passengers.

Reintroduction of School Bus Service

Until the early 1980s, Colombo City had a reasonably well-patronized school bus service operated by the state-owned Sri Lanka Transport Board. However, with the introduction of private buses and the 'peoplization' of the state-owned bus operations, school bus services have been neglected in favour of more remunerative services. This vacuum has been largely filled by private vans that offer a charter service for a monthly fee. There are presently only some 50 to 60 school buses operating to and from Colombo City, compared to an estimated 2,167 vans (having increased from 1,619 vans in 1996).[58] The breakdown of the arrivals by main corridor estimated for the year 2001 at a 5 per cent per annum growth rate since 1996 is given in Table 6.20. There have been on average 16 students per van.

Table 6.20 Number of school vans arriving in Colombo City (estimate for 2001)

Corridor	Vans
Galle Road–Dehiwela Bridge	395
Horana Road–Pamankade Bridge	221
High Level Road–Kirullapone Bridge	220
Negombo Road–SLJF Bridge	292
Kandy Road–New Kelani Bridge	505
Low Level Road–Orugodawatte	9
Dematagoda Road	82
Parliament Road–Castle Street	316
Cotta Road–Borella	59
Nawala Road	68
Total	2,167

These school vans have caused a number of traffic and parking related problems that have drawn the attention of policy-makers in the transport sector. Presently, there is an attempt to reintroduce school buses on a large scale. Discussions with school principals have revealed that approximately 50 per cent of the students (and teachers) travelling in these vans would transfer to a regular and dependable school bus service and that each bus would carry on average 80 students. Thus each bus would replace five school vans, with a saving of at least 50 per cent in road space.

The economics of the present school van service are opaque.[59] The 1996 survey showed that the average cost of school van service was Rs 570/month per student, only slightly lower than that charged by office vans (Rs 650/month).[60] The survey also suggested that most school vans were used only for this purpose. In the case of the school buses run by public bus companies, students are charged 10 per cent of the normal fare; government provides a 55 per cent subsidy; and the balance of 35 per cent is covered by the operator.[61]

Table 6.21 shows the benefits and costs of this policy option, expressed as NPVs over an assumed 20-year time horizon. Environmental benefits account for some 7 per cent of the total economic benefits.

Table 6.21 Benefits and costs, NPV, million Rs

	Road users	Students	Govt	Total	Local Env	Global Env	Total
VOC savings	736			736			736
Time savings	512	x		512			512
Fuel taxes	30		-30	0			0
Fees	x	x	x	0			0
Environment							
PM-10				0	13		13
NOx				0	25		25
SO_2				0	24		24
GHG				0		33	33
Total	*1,277*		*-30*	*1,247*	*61*	*33*	*1,342*

Notes: positive entries=benefit; negative entry = cost
x = undetermined
Fuel savings are included in VOC savings

The distributional impacts are difficult to disaggregate. All other things being equal, it would be reasonable to assume that if one bus replaces five vans, that some reduction in both the government subsidy and the fees charged to students would be achievable.[62] Precisely how the (economic) vehicle operating cost savings would be distributed among bus operators, students and government is difficult to say. We also show as undetermined the time savings to the students themselves: these may be hard to express in monetary terms (since, for the vast majority, there is no opportunity cost implied from lost earnings); however, there is certainly some gain in utility.

Even if one were to make the pessimistic assumption that the reintroduction of school bus service would require brand new buses (and that the replaced vans have zero salvage value), at an estimated cost of Rs 860 million, the net benefits would still be Rs 482 million. Thus the reintroduction of school buses is self-evidently win–win, and is therefore strongly recommended.

NOTES

1 As noted by Jayaweera (1999), every large congested city harbours a reservoir of potential peak period drivers (latent demand) – who are deterred from adding to congestion only by congestion itself. Once a new highway eases congestion, these drivers, who may now be travelling in other modes, or at other times of the day – will move into the empty spaces, thus driving congestion levels back up.

2 Jayaweera (1999) notes the distinction between road pricing and congestion pricing as follows: road pricing refers to a well-defined charge for particular trips at well specified places or times – conventional road tolls being the typical example (and in general unrelated to the specific problem of congestion externalities). On the other hand, congestion pricing refers to schemes in which the price charged is targeted to the users who cause the greatest problem – namely those who travel during already congested peak hours.

3 Peak hour speeds in Singapore are typically 25 km/hr during the evening peak, and 30 km/hr in the morning peak, compared to speeds of 10 km/hr in New York, 12 km/hr in London, and less than 10 km/hr in Bangkok (Chin, 2000).

4 The international experience with congestion pricing, including that of Hong Kong and Singapore, as well as several European cities, is reviewed in Jayaweera (1999).

5 This represents about 1.5 per cent of the 1996 GDP of Rs 768 billion.

6 These are noted by Chandrasiri (1999, p.10). For example, special privileges are extended to state corporations such as the Sri Lanka Central Transport Board (SLCTB), formerly the state monopoly. Special permit schemes for public servants and politicians were terminated in 1994, then reintroduced in 1998, and again rescinded in March 2001.

7 However, there are four companies (Dimo, Latec, Ceymo and Ashok Leyland) which assemble trucks and buses in Sri Lanka. The assembly operation involves fixing of body kits and chassis imported from four major sources: the UK, Japan, India and China.

8 That is, cif plus all duties and taxes, but excluding dealer margins.

9 Nominal cif prices in US$ would of course have increased over this period, so the general trend would expected to be upward.

10 For 2000, we estimate only some 17,000 diesel passenger cars in the fleet, as opposed to 130,000 gasoline cars, and 158,000 diesel "dual-purpose" vehicles.

11 In all cases, except where noted, the dependent variable is consumption per vehicle. We note that several US studies use consumption per driver – see, for example, Greene (1999).

12 The estimates by Meier et al. (1997) do include the early years of liberalization, but their time series is still relatively short and extends only to 1991. They used the so-called vehicle use lagged endogenous model, which models demand as the autoregressive function Q(t) = f (Q(t-1), P, Y, ...), and which allows separate estimation of long- and short-run elasticities.

13 For example, McRae (1994), who examined aggregate gasoline demand in several developing Asian countries, does not once mention motorcycles or three-wheelers.

14 McRae (1994) notes "more vehicles per capita are expected to reduce utilisation, so the expected sign for the variable is negative".

15 In this technique (SURE = seemingly unrelated regression equations), equations are estimated simultaneously to avoid problems when the error term across equations are correlated, which is the case in fuel demand equations linked by cross-price elasticities. For details see, for example, Lutkepohl (1991).

16 This is the form used by Meier et al., (1997) and Jayaweera (1999).

17 Alternatively, one could adjust the number of vehicles to automobile equivalents (in the gasoline equations), or heavy vehicle equivalents (in the diesel equation)

18 The elasticities were estimated using data for 1988 to 2000 and estimated in log-linear form using OLS.

19 Some discussions of the issue, for example, Chin (2000) and Jayaweera (1999), use the term "average variable cost, AVC" for this quantity; we prefer the term average private cost (APC) to stress the difference between private and social costs.

20 Following Pigou's classic study that laid the theoretical foundation for road congestion pricing (Pigou, 1924).

21 In some cases, it is the individual's personal opportunity cost of time that would decide whether he or she would pay the expressway toll in exchange for the time savings (or whether a bus passenger would choose expressway service rather than a slower stopping service on the A30 in exchange for a slightly higher fare). How that opportunity cost is related to an individual's salary is an empirical question, about which there are, in Sri Lanka, no data (since there are no toll roads in operation that would allow estimation of the trade-offs made by users). Absent such evidence, average salary is surely the best guide.

In the case of drivers, the individual's valuation is irrelevant, for as (mostly) salaried individuals, they may be assumed to be indifferent to the speed at which their vehicle travels. However, their employers would not be indifferent: and it is reasonable to assume that the opportunity cost they place on driver time is equal to salary (plus fringe benefits and employment taxes).

In the case of bus passengers, again largely salaried commuters, the consequence of road congestion is longer commute times, which reduces leisure time (or sleep) rather than hours on the job. Again there is no data for what opportunity costs such passengers place on their leisure time, but it is not unreasonable to assume it would be related to income.

22 A March 1996 survey showed that bus passengers (Rs 2,900) had significantly lower average incomes than both suburban rail passengers (Rs 4,900) or office van passengers (Rs 4,900) (Jayaweera 1999). The TSSS deduces that this unexpected result is a consequence of the fact that a majority of Colombo suburban rail passengers are civil servants, whose employing departments pay 75 per cent of the season ticket cost.

23 Assuming 22 working days per month and 8 hours/day.

24 Which corresponds exactly to the figure given in Table 6.8 for monthly car passenger salary. This is the figure used by the University of Moratuwa study.

25 Clearly the fraction of car passengers whose trips are part of their business activity is less than one. Indeed, with the advent of the cellular phone, much business is conducted from automobiles, necessarily implying a reduction of the opportunity cost of being stuck in congested traffic.

26 The older literature is reviewed in a 1987 report by the UK Institute of Transport Studies (The value of travel time savings), which noted, inter alia, that "higher absolute values of time are associated with higher levels of income . . . there thus appears to be some justification for custom in benefit cost studies using values for time for future saving which are adjusted for future changes in disposable income." See also Jayaweera (1999), pp. 44–6.

27 This is comparable to the time value per hour for the vehicle as a whole (that is, passenger(s) + driver) of Rs 66/hour used in the CKE-EIS. It might be argued that higher values are justified for the expressway/A30 corridor, since the proportion of higher-valued business trips is likely to be higher than the Colombo averages as determined by the Univeristy of Moratuwa studies.

28 This proposal anticipates a monthly sticker of Rs 200, based on 25 days/month, that is, Rs 8/day (though the proposal states the daily toll is Rs 10 'with a 20 per cent concession').

29 This uncertainty is not limited to Colombo: the claim that the proposed daily congestion charge for central London of £5 would reduce traffic by 15 per cent was highly controversial at the time the charge was being proposed. More than a year after the charge came into operation, peak traffic in central London has undoubtedly diminished, but how much of the decline is attributable to the charge, and how much to the business cycle (and the lower number of tourists in 2003 following the Iraq war, which started at the same time as the charge was introduced), is still unclear. However, one measure of the success of the London congestion charge is the eagerness of other congested UK cities to introduce similar schemes.

30 It is evident that the calculation of expected revenue by the Colombo Municipal Council does not take into effect any reduction of traffic due to the elasticity effect – indeed, from the Council's perspective the proposal is entirely one of raising revenue for road maintenance.

31 Assuming average occupancies of 28 passengers in medium buses, and 45 in large buses.

32 It is possible that some people would also take the train rather than buses, though we ignore this potential effect since we have no data to estimate it.

33 Indeed, although the buses themselves experience the same time savings as the cars consequent to higher speeds, because of the general need to wait for buses at bus stops, and the time involved in walking from home to the bus stop, and from the Colombo bus stop to the place of work, there may be no net time saving for the person who shifts from car or motorcycle to bus. Since the number of car passengers who shift from cars to buses is small compared to the number of existing bus passengers, we simply assume such additional utility losses are captured in the (simple) calculation depicted in Figure 6.12.

It is true that bus fares are certainly at significantly lower levels than the corresponding costs of a car trip, but this is self-evidently not a consideration for those who can afford the comfort of automobiles (as opposed to what is perceived as the discomfort of buses). The direct savings (i.e. actual private VOC v. bus fare) are likely to be much smaller than the value assigned by such individuals to their time; therefore this (small) gain can also safely be ignored in the calculations of aggregate economic benefits.

34 However, the lack of substantive merit is rarely a reason for vested interests not to protest. The impact of the price rise of diesel fuel is discussed further in Chapter 7 in relation to fuel tax equalization.

35 The daily congestion charge shown is for passenger cars: tolls for other vehicles are scaled in proportion to the values shown in Table 6.12 (for example, when the passenger car toll is Rs 25, the motorcycle toll is Rs 10; therefore, when the passenger car toll is set at Rs 50, the motorcycle toll is set at Rs 20, and so on).

36 The first report appears to have been prepared in 1984 by the Japan International Cooperation Agency.

37 For a scathing critique of the original environmental impact study, see, for example, Sivagnasothy and McCauley (2000). However, it should be noted that this critique was based on the earlier 1993 environmental impact analysis (Road Development Authority, 1997). The subsequent report of June 1997 examined many of the issues previously omitted. Unlike the original analysis, it also included a monetization of the benefits of reduced congestion on the A3 and the reduced costs of accidents.

38 However, it is reported that some significant unforeseen geological conditions were encountered soon after work began, and that some significant further delays are now anticipated.

39 Three major studies of the Southern Highway project have been conducted, the most recent of which is Wilbur Smith Associates (1998). These studies were reviewed by Kumarage (1999). None of the three studies examined the impact on vehicle emissions.

40 It is not only those vehicles who elect to use the new expressway that derive benefits from reduced journey times, but, *ceteribus paribus*, users who continue to use the old A30 will also derive benefits because traffic volumes will be lower, and traffic speeds higher (since some fraction of the existing traffic will now be diverted to the expressway). These effects are captured in the model, though not illustrated in the figure.

41 The coefficient 0.5 in Equation (6.4) assumes a linear demand curve (as drawn in Figure 6.15). Depending upon the elasticity, the area EFG may be much less than half the product $\Delta C\ \Delta G_1$.

42 In other words, we assume that $\Delta C^* \sim \Delta B_2 - \Delta B_1$.

43 This has been used in many recent transportation planning studies in Sri Lanka. For a description, see A. Kumarage et al., 2000.

44 University of Moratuwa (2000), section 5.5 notes that an increase of 50 per cent [for the passenger vehicle toll] results in an decrease of 10 per cent in traffic count. Therefore $e = [dQ/Q]/[dP/P] = -0.1/0.5 = -0.2$.

45 These are for scenarios #1 and #8, respectively. Scenario #1 is based on a 3 per cent GDP growth rate, scenario #8 on 6 per cent.

46 Interestingly, Sigvagnasothy and McCauley chose not to value this: "the other potentially significant negative environmental impact identified but not valued was the loss of biologically diverse wetlands, but it did not appear that the route chosen for the Expressway would likely result in a major loss of such eco-systems" (p. 120).

47 Undoubtedly some effort would be made to charge higher fares for buses using the Expressway – though since the toll is far smaller than the time and fuel savings, the argument that the bus operators would incur higher costs in total appears to have little merit.

48 There appear to be no detailed studies on the question of the incremental electricity consumption of the EMUs. The fuel consumption of a DMU has been estimated at 2.5 litres/km (Kumarage, 1999); we present below the calculations that convert the corresponding energy requirement to coal inputs at the (remote) coal-based power station. The assumption that coal would be the power generation source is conservative, since PM-10 emissions from an oil-fired (steam-cycle) power station would likely be lower. The use of LNG would of course be the most environmentally desirable form of power generation (see Chapter 5), but its costs (at the scale necessarily implied for Sri Lanka) are prohibitive (even assuming it were technically feasible, about which there are serious questions). Of course, given the prolonged paralysis over the coal plant, it may well be that the most likely source of thermal power would be short-gestation diesel plants, but this too is a very expensive way of providing electricity for railway electrification.

49 Value obtained from Kumarage (1999).

50 It might be argued that the DMUs that are presently in service on the main line, and which would be replaced by EMUs, would be available for use elsewhere, which would induce additional bus passengers to transfer to DMUs on these routes (for example, the coastal line) in response to the better service that would be offered. If that were the case, however, then one would also have to add to the analysis (i) the O&M costs of that additional DMU based service; (ii) the emissions associated with that service (which

would be roughly equal to those of the present service, but which in turn would mean we would also need to add the emissions from electricity generation). This would have made the analysis much more complicated, and we therefore elected to simply consider a credit (salvage value) of the DMUs that would be replaced by EMUs.

51 Based on load factors given in Ministry of Transport and Highways, Bus Transport Policy 1999.

52 DMU units would more accurately be described as "diesel-electric multiple units", because the diesel is used to generate electricity which in turn powers axle mounted electric motors. Thermodynamically this is equivalent to the EMU, except that the electricity is not generated on board the unit, but at a remote location. While in theory one would need to account for transmission and voltage transformation losses in the EMU, the efficiency of large scale generation would be better (even if generated by the same diesel technology), which offsets these losses.

53 4.6 million litres = 3,840 tonnes of diesel. Ignoring EHV transmission losses, this is equal to 7,434 tonnes of coal (based on a calorific value of coal at 6,300 Kcal/kg and 10,180 Kcal/kg for diesel oil). Assuming carbon fractions of 0.875 for diesel and 0.6 for coal, it is clear that a shift from diesel fuel to coal would increase GHG emissions (from 3,365 tonnes year to 4,460 tonnes/year).

54 Clearly this is the expected finding: it does not matter where carbon emissions are produced, since, unlike the local air emissions, they all have the same (worldwide) damage cost.

55 Including a 1993 World Bank study of container traffic; Stage 1 of the Colombo Urban Transportation Study; and a 1996 JICA study of the Development of a New Port for Colombo.

56 For which we assume trip lengths of 12 km and 6 km, respectively.

57 Thus the value of freight delay of Rs 150 per vehicle-hour, noted earlier, is reduced to Rs75/hour.

58 Study of Transportation Services Provided by School and Office Vans, Transport Studies and Planning Centre, May 1996 (published in World Bank (1996b), Volume II, pp 146–202). This earlier study was limited to gathering physical data, and presented no benefit–cost analysis (or quantitative estimates of the impact on reducing road congestion).

59 School and office vans are eligible for concessionary diesel tax, but this requires production of a certificate that the van has been registered as a school van, but operators are not compelled by law to insure passengers (World Bank, 1996b).

60 World Bank (1996b, p. 150). This is significantly higher than the corresponding bus fare. If we take Rs 5.50 as the normal fare (without transfer) for a 10 km trip, then the average bus fare cost would be some Rs 240/month.

61 It is this apparent loss that explains the reluctance of bus companies to provide school bus services: though why under these conditions any bus service would remain at all is unclear.

62 The costs of the present van service constitute a significant burden on many parents. While there is no systematic data, Kumarage (2000) gives an example of a family who spend Rs 400/month to send their daughter to school by van because of the poor bus service, compared to a monthly daily commuting bill of Rs 800/month. For this distance (in mid-2000), the normal bus fare would have been Rs 5.50 per trip, or Rs 5.50 x 22 x 2 = Rs 242/month, so the van cost is substantially greater.

7. The economic and environmental costs of fuel price distortions*

In this chapter we examine the consequences of distortions in fuel prices. We begin with what is an almost classical example: the conversion of gasoline passenger cars to LPG – for which the motivation is entirely the subsidy on LPG. We then examine the effect of the discrepancy in tax rates on gasoline and diesel fuel, and examine the impacts of a tax equalization.

7.1 CONVERSION OF PASSENGER CARS TO LPG

Over the past 20 years, LPG use in Sri Lanka is the energy form that has seen by far the highest growth rate of consumption, averaging some 16 per cent per year, as opposed to 7 per cent per year for diesel fuel, and about 7 per cent for electricity. Since refinery production is quite limited, this has resulted in an exponential growth of LPG imports, as shown in Figure 7.1. Much of the growth in the late 1990s was due to the conversion of gasoline cars to LPG: we estimate that by the end of 2000 some 15,500 gasoline cars and dual-purpose vehicles – about 10 per cent of the fleet – had been converted.[1] However, following LPG price increases in 2000, the rate of growth of LPG consumption appears to have levelled off.

Conversion of vehicles to LPG is attractive from an environmental point of view, since LPG is among the cleanest transportation fuels, comparable to ethanol and compressed natural gas (CNG). However, there are two disadvantages: LPG is relatively expensive, and, for safe operation, vehicle conversion needs to be subject to strict regulation. Although the Government is studying ways of legalizing the use of LPG as a motor fuel, so far no rules and regulations have been gazetted by the Commissioner of Motor Traffic.

* Chapter 7 contributor: S. Chandrasekera.

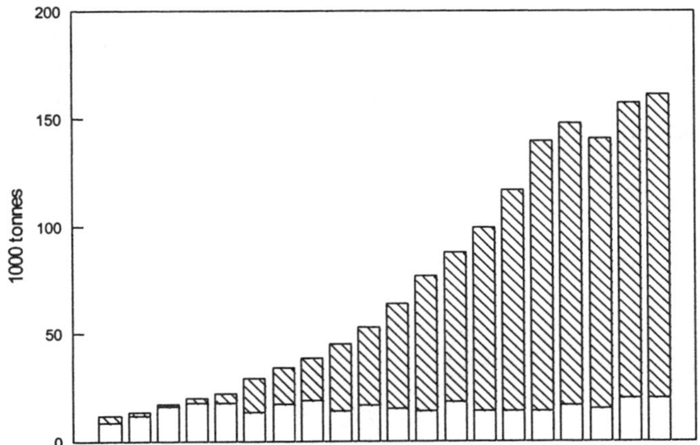

	84		86		88		90		92		94		96		98		00		02	
refinery	9	12	16	18	18	13	17	19	14	17	15	14	18	14	14	14	17	15	20	20
imports	3	2	1	2	5	16	17	19	31	36	49	63	70	85	103	126	131	126	137	141

Figure 7.1 Growth in LPG consumption (1000 tonnes, by source)

There is much anecdotal evidence – though few firm numbers – about the extent to which gasoline automobiles have been converted to LPG over the past few years. Some cars (and particularly taxis) use LPG in domestic bottles, while others are equipped to make use of official LPG outlets, all in Colombo. Vehicle conversion costs also differ considerably: properly designed on-board storage tanks cost 40,000–50,000 Rs/vehicle, as compared to conversion costs of Rs 15,000–20,000 for using domestic bottles. Data on how many vehicles have been converted to use domestic bottles is particularly hard to estimate (since their owners are understandably reluctant to admit to this type of conversion).

Table 7.1 Annual average differentials

	fob US$/tonne	cif US$/tonne
1997	25.35	95.35
1998	5.25	75.25
1999	24.75	89.05
2000	35.90	82.60
2001 (to June)	46.90	89.10

Figure 7.2 shows the difference between fob prices for gasoline and LPG. The differential is strongly seasonal, with a maximum value in the winter months. Table 7.1 shows the annual averages, with cif prices as shown in Figure 7.3. Cif differentials are higher than fob differentials because the costs of transportation to Colombo are much higher for LPG.

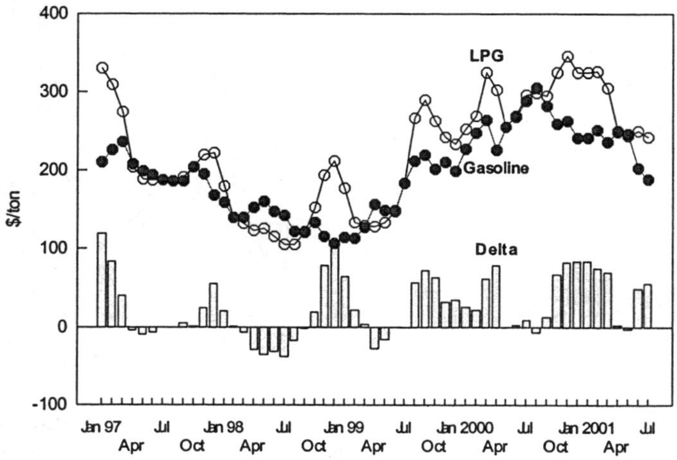

Notes: LPG: Aramco posted price, ex-Gulf
Gasoline: Mogas RON 92, Singapore spot, *Platt's Oilgram*

Figure 7.2 FOB prices, gasoline and LPG ($/tonne)

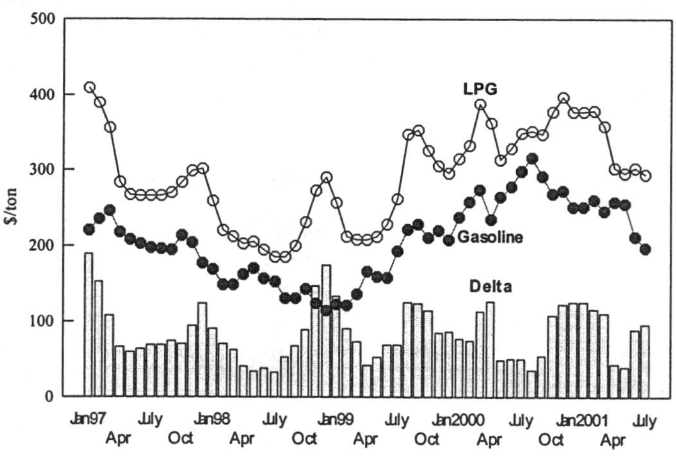

Figure 7.3 CIF prices, gasoline and LPG ($/tonne)

Finally, Figure 7.4 shows a comparison of the delivered price at the pump as Rs/litre: throughout this time the actual retail gasoline price of Rs 50/litre was over triple the economic price, while the LPG price was between Rs 26.50 and 29.50/litre. When the difference in density is taken into account, LPG is about 30% cheaper.

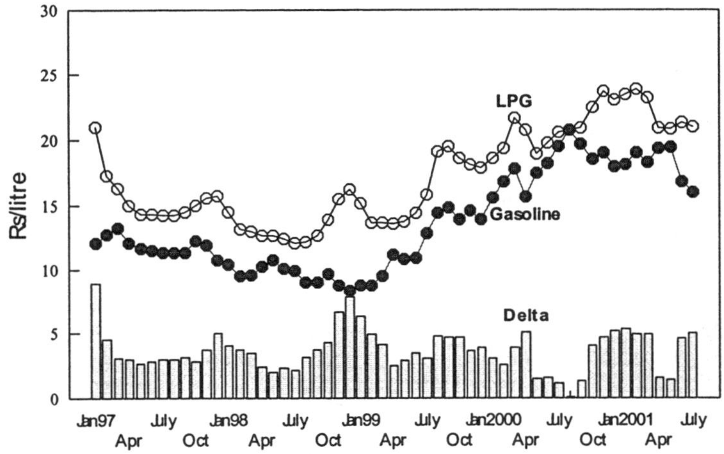

Figure 7.4 Delivered economic cost, Rs/litre

The delivered cost of LPG is strongly dependent upon transportation modality and cost. Table 7.2 shows the price structure under various transportation modes. The key assumptions are as follows:

- Tanker size: the ARAMCO terminal normally handles refrigerated vessels with capacity greater than 13,000 metric tonnes, for which freight to Sri Lanka is US$ 35–45 per tonne. For the smaller pressurised cargo vessels, available in standardised 1500, 2500, 4000, and 5000 tonne sizes, freight may be as high as US$ 100/tonne. The freight paid by local suppliers is reported to have come down from a high of US$ 80 per tonne in early 1999 to US$ 52.25 in June 2000, possibly due to larger parcels imported. LPG could also be imported in tanktainers of eight metric tonne capacity, for which the average freight may be around US$ 25/tonne. Tanktainers are pressurized containers that can be unloaded directly onto a suitable road vehicle and delivered without transhipment to the (bulk) consumer (such as a filling station). This is a new concept for LPG that is not yet widely practised, but it has great potential to reduce delivered costs.
- Terminal costs: most terminal operators state that they would plan for a margin of US$ 100 (between input and output prices). It is reported

that operators presently are compelled to work with a lower margin of around US$ 50 per tonne.

- Local transportation: the present local charges for bulk deliveries vary according to the parcel size. For parcel sizes up to 2,000 kg, Rs 12/kg per km; for parcel sizes of 2,000 to 4,000 kg, Rs 0.10/kg per km; and for parcel sizes above 4,000 kg, Rs 0.09/kg per km.

Table 7.2 Cost of LPG

	ARAMCO price (US$/ton fob)	Freight per ton in US$	National Security Levy (c)	Terminal cost (US$/ton)	Total ex-terminal (Rs/kg)	Delivered cost (city) (Rs/litre) (b)
Refrigerated 13 000 tonnes	250.00	40.00	23.56	100.00	33.91	18.65
Pressurised: 1 500 tonnes	250.00	100.00	28.43	100.00	39.23	21.57
2 500 tonnes	250.00	80.00	26.81	100.00	37.45	20.59
5 000 tonnes	250.00	52.25	24.57	100.00	34.99	19.24
1 500 tonnes	250.00	100.00	28.43	60.00	35.95	19.77
1 500 tonnes	300.00	52.25	28.64	100.00	39.43	21.68
Tanktainer	250.00	25.00	22.34	10.00(a)	25.20	13.86

Notes:

(a) Local handling cost.

(b) Present selling price of LPG at Auto Gas outlet Rs 26.50 per litre in the city and Rs 29.50 in the outstations; Present delivered price of LPG to Auto Gas dealers Rs 34.00 per kg. Transport charges: Rs 0.10/kg/km.

(c) National Security levy = 6.5 per cent.

Exchange rate US$ = Rs 82.00;

Conversion factor: Litre LPG = 0.55 kg

Table 7.3 Emission factors (gm/km)

		Gasoline	LPG
Monetized	NOx	2.19	2.16
	PM-10	~0.00	~0.00
	SO_2	~0.00	~0.00
	Lead	0.15	0.00
Not monetized	CH_4	0.08	0.056
	VOC	4.63	1.51
	CO	49.97	7.17
	N_2O	0.005	0.00

Source: Chandrasiri (1998), Annex 4

Table 7.3 compares the emission factors for gasoline vehicles using gasoline and LPG. Among the emissions we monetize, only for lead is there a significant difference. However, LPG vehicles do provide an 80 per cent emission reduction for CO and VOC.

Table 7.4 shows the benefits and costs of LPG conversion. We assume that LPG conversion continues at about 5 per cent per year, maintaining the present 10 per cent of the petrol-engine vehicle fleet. Some arithmetic confirms that at present price levels, the terminal margins are around US$ 50/tonne, as assumed here.

Table 7.4 Costs and benefits of LPG conversion (v. leaded gasoline) Rs million (NPV at 10 per cent discount rate over 20 years)

	Road users	Govt	Total	Environ-ment	Total
Vehicle conversion costs (a)	-188		-188		-188
LPG, economic cost	-15,457		-15,457		-15,457
Gasoline, economic savings	12,431		12,431		12,431
Gasoline taxes	16,687	-16,687	0		0
LPG taxes (NSL)	-1,431	1,431	0		0
Emissions reductions					
Lead			0	4,077	4,077
VOC, CO			0	(b)	0
Total	12,042	-15,256	-3,213	4,077	864

Notes:
(a) Assuming conversion cost of Rs 20,000/vehicle (for bottled gas).
(b) Not monetized.

LPG conversion brings significant environmental benefits only if it displaces leaded gasoline; if leaded gasoline is phased out, the environmental benefits are marginal, and do not offset the economic costs of vehicle conversion and the higher cif costs of LPG imports. Clearly road users obtain significant (financial) benefits, but these are solely a consequence of the distortions in taxation.

Implementation of the tanktainer concept would significantly lower the economic costs, for the freight costs reduce from the present US$ 50/tonne to US$ 25/tonne, and terminal handling costs from US$ 100 to US$ 10/tonne – for a total of US$ 35/tonne (as opposed to US$ 10/tonne for gasoline). However, the delivered economic cost at the pump would still be higher than that of gasoline, so the economic benefits before consideration of the environment are still negative at Rs -1,315 million, (though reduced from the Rs -3,213 million of Table 7.4).

Table 7.5 Costs and benefits of LPG (tanktainer concept)

	Road Users	Govt	Total	Environ-ment	Total
Vehicle conversion costs	-422		-422		-422
LPG, economic cost	-13,324		-13,324		-13,324
Gasoline, economic savings	12,431		12,431		12,431
Gasoline taxes	16,687	-16,687	0		0
LPG taxes (NSL)	-1,329	1,329	0		0
Emissions					
Lead			0	4,077	4,077
VOC, CO			0		0
Total	14,043	-15,358	-1,315	4,077	2,762

We draw several conclusions relevant for transport sector policy:

- With 10 per cent of gasoline vehicles presently converted to LPG there is a significant loss of gasoline tax revenue to the government. More important, with as many as 7 per cent of gasoline engine vehicles (that is, some 10,000 vehicles) likely using domestic bottles, any policy instigated now to ban use of these in cars is not likely to be practical. Thus, initially, the emphasis should be on an inspection regime to ensure safe installation and operation.
- The conversion of gasoline vehicles to LPG is driven entirely by distortions in tax policy, and the resulting price differential at the pump between highly taxed gasoline, and LPG (subject only to the National Security Levy). In effect, the present price structure provides a significant tax exemption to owners of vehicles that have above average annual vehicle-km, notably taxi fleets.
- The effective tax exemption for domestic LPG also benefits the same upper income groups who also own gasoline passenger cars and use the car-taxi service:[2] the overwhelming majority of the poor still use fuelwood for cooking, and would be affected little by increases in LPG prices.

Box 7.1: LPG subsidies in Yemen

Subsidies on petroleum products are encountered throughout the world. Everywhere subsidies have noble intentions, but rarely do they meet their main goal to assist the poor.

Yemen is an excellent example of the failure of diesel subsidies to reach the poor (World Bank, 2005). In 2003, the economic price of diesel was 39.8 YR/litre (US 22 cents/litre), as opposed to an actual retail price of 17 YR/litre, a subsidy of over 50%. The cost to the Government was YR 56 billion (US$ 300 million) for diesel and over YR 100 billion for all petroleum products (including LPG) – equal to about 65% of all Government development expenditures. Yet little of this subsidy actually reaches the poor: as shown in the figure, the bottom decile income decile (d[1]) captures only 1% of the subsidy (associated with direct purchases of diesel by households), while the top income decile (d[10]) captures 40%.

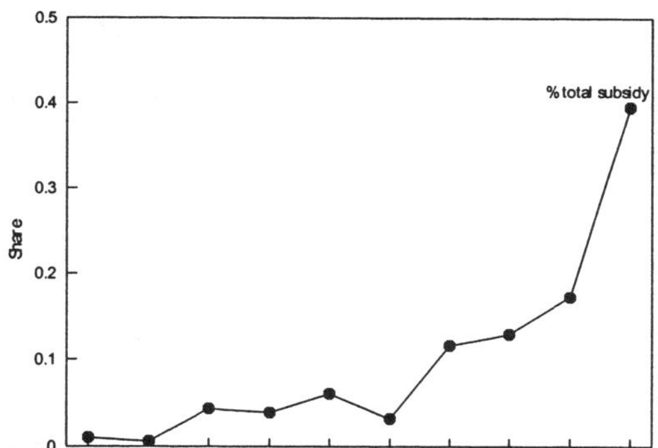

	d[1]	d[2]	d[3]	d[4]	d[5]	d[6]	d[7]	d[8]	d[9]	d[10]
● % total subsidy	1%	1%	4%	4%	6%	3%	12%	13%	17%	40%

This reflects the fact that most of the diesel purchased by households is bought by wealthy farmers for irrigation pumping (and particularly those who have become wealthy by growing Qat, the most profitable cash crop in Yemen). Indeed, the highly subsidized diesel price is one of the reasons for the rapid depletion of the groundwater resource and rapidly increasing depths of the groundwater table. This in turn imposes higher costs on all for drinking water and food production.

However, the poor do benefit from the indirect effects of the subsidy (subsidized diesel reduces the production and transportation cost of food), and concern over potential food price increases is constraining the Government's desire to reduce subsidies. However, by far the best solution is for diesel to be priced at its economic cost, and for Government to use the corresponding increase in its revenue for social spending, including direct cash grants to the poor through the Social Welfare Fund – though this is far from easy, given the difficulties of identifying the poorest segments in outlying rural areas.

7.2 TAX EQUALIZATION ON GASOLINE AND DIESEL FUEL

Historical tax differentials on diesel and gasoline (both on fuel and on vehicle import duties) have had a seriously distorting impact on the fleet mix, as well as on the economics of the refinery. High taxes levied on gasoline have resulted in a relatively low gasoline demand, with the result that, for several decades, naphtha has had to be exported (at a loss), while diesel has had to be imported.[3] As shown in Figure 7.5, the ratio of gasoline to diesel price has varied over the past 30 years from 3.5 to 1.5. This may be compared to rough equivalence of Singapore prices. Table 7.6 shows recent prices for petroleum products.

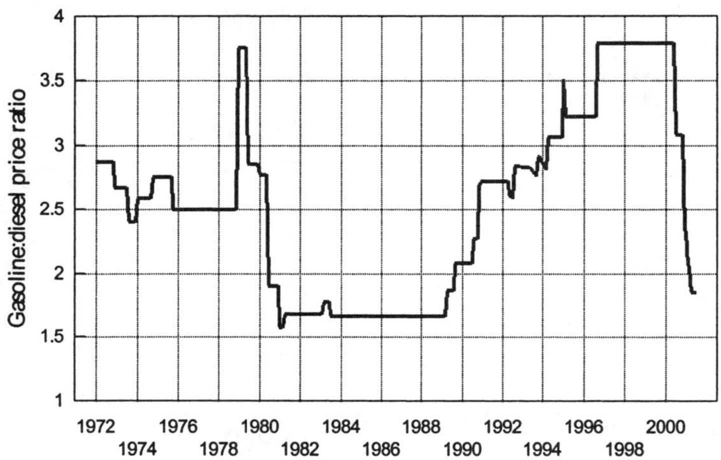

Figure 7.5 Historical ratio of gasoline to diesel price

Table 7.6 Retail petroleum product prices (year end, Rs/litre)

	2001	2002	2003
Gasoline (90 Octane)	50.00	49.00	53.00
Gasoline (95 Octane)	53.00	52.00	56.00
Auto diesel	26.50	30.00	32.00
Kerosene	17.40	24.00	25.50
Furnace oil (1,000 seconds)	16.50	21.20	23.20
LPG (Rs/kg): Shell gas	40.72	47.84	45.60

The earlier *ad hoc* petroleum pricing system has now been changed to allow more regular adjustment to international price levels: the new system introduced in 2002 is illustrated in Table 7.7, and applies to kerosene, gasoline, diesel and fuel oil. In principle, this formula is applied to the Singapore Platts quote every month; for prices at the pump to be revised requires a minimum increment of Rs 0.25/litre, and is also subject to a maximum monthly increment of Rs 2.00/litre.

Table 7.7 Current price structure

	$/bbl	Petrol	Diesel	
World oil price	$/bbl	35.00	35.00	
fob Singapore [Platts]	$/bbl	46.22	44.20	average of the previous month
	bbls/ton	8.38	7.52	
Freight $/ton	$/ton	11.00	11.00	
Jetty pipeline charge	$/ton	3.00	3.00	
Freight&jetty	$/bbl	1.67	1.86	
cif, $/bbl	$/bbl	47.89	46.08	
Exchange rate	[Rs/$]	94.83	94.83	average of the previous month
cif, Rs/litre	Rs/litre	28.58	27.50	
Stamp duty	Rs/litre	0.34	0.33	
Landed cost	Rs/litre	28.92	27.83	
Excise duty	Rs/litre	30.00	6.50	fixed amount for 2002 and 03
Customs duty	Rs/litre			
	Rs/litre	58.92	34.33	
NSL	Rs/litre	2.32	2.23	
GST (now VAT)	Rs/litre			presently only on fuel oil
Tax paid landed cost	Rs/litre	61.25	36.57	
CPC finance charge	Rs/litre	2.54	2.54	fixed amounts for 2002
CPC wholesale costs	Rs/litre	0.24	0.24	for 2002 only
CPC margin	Rs/litre			for 2002 zero on petrol and diesel
CPC wholesale cost	Rs/litre	64.03	39.35	
Marketing+distribution	Rs/litre	1.23	1.23	for 2002 only
Dealer discount	Rs/litre	0.46	0.26	
Consumer retail price	Rs/litre	65.72	40.84	

However, this pricing system has run into trouble in 2004, as the government has been reluctant to pass soaring world oil price increases to the consumer: just for the first six months of 2004, the Government must foot a bill of some Rs 6.5 billion to CPC and Rs 1.5 billion to LIOC. Table 7.7 shows the working of the system for a world oil price of US$ 35/bbl, which results in prices of Rs 66/litre and Rs 41/litre for gasoline and diesel, respectively. In July 2004 the government finally raised the gasoline price to Rs 65/litre (bringing the gasoline price in line with a world crude price of US$ 35/bbl), but diesel and kerosene prices remain unchanged. Thus the effective tax differential between diesel and gasoline remains as wide as ever.

As in many other countries, the ostensible rationale for the high level of tax on gasoline is that gasoline is mainly a fuel used by the those with relatively high incomes, while diesel is mainly a fuel used by those with relatively low incomes. That may have been the case before liberalisation, when car ownership was indeed limited to a small segment of the population, and diesel vehicles limited to buses, trucks and tractors. But with the dramatic changes in vehicle mix, such a rationale, as elsewhere, can no longer be sustained – as is readily evident from a breakdown of gasoline consumption by vehicle type (Table 7.8). Gasoline cars account for only 36.5 per cent of total gasoline sales (and hence of gasoline taxes), while motorcycles and 3-wheelers account for 56 per cent, self-evidently the personal transportation mode used by lower income workers.

Table 7.8 Breakdown of gasoline use by vehicle class, 2000

	1000 tons	[%]
Autos	72	36.5
Vans	10	
Land vehicles	1	
Motorcycles	59	30
3-wheelers	55	28
Total	197	

Data on vehicle ownership by income distribution are instructive: Figure 7.6 shows the distribution of vehicle ownership by income class. Motorcycles are the dominant vehicles for 98 per cent of earning households; only in the top 2 per cent (earning more than Rs 25,000/month) are cars the dominant vehicle.

In sum, there is no evidence that the present tax structure on transportation fuels is progressive. In fact the present tax structure disproportionately favours the better off – most importantly the owners of so-called dual-purpose vehicles, whose consumption of diesel just in this category (232,000 tonnes in 2000) exceeds the total consumption of petrol.

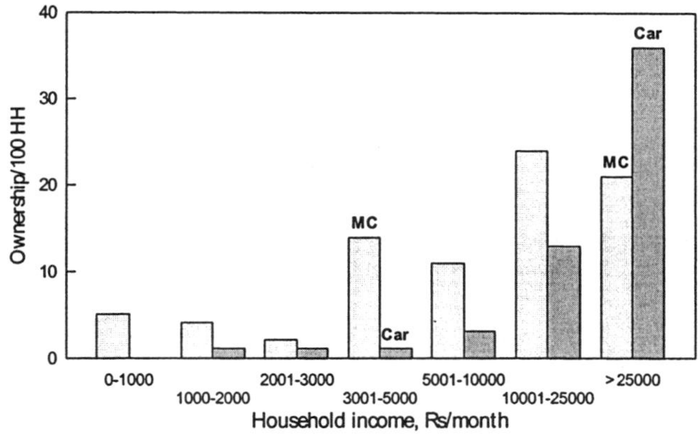

Figure 7.6 Vehicle ownership per 100 households by income

In addition to the distributional rationale for equalisation of diesel and gasoline taxes, there is also an environmental rationale. In the case of petrol, the main local air emission of concern in Sri Lanka is lead and particulates from 2-stroke engines – emissions that would increase in any tax-induced shift to gasoline vehicles. But these issues are more easily solved (by introducing unleaded gasoline, as discussed in Chapter 8, or limiting 2S-motorcycles) than reducing the high emissions of particulates from heavy diesel vehicles.

Finally, there is the further rationale of better matching road user charges to public costs for road maintenance. As noted by Jayaweera (2000), two-axle trucks have the heaviest axle loads, and contribute most to road damage: as shown in Table 7.9, a two-axle medium truck (the most common type in Sri Lanka) imposes road maintenance costs of Rs 7/vehicle-km, as opposed to Rs 1.35/vehicle km for passenger cars. It is evident that while the gasoline tax on passenger cars is 210 per cent of the road maintenance costs imposed, the diesel fuel tax paid by trucks is between 6 per cent and 22 per cent. But trucks are exempt from the special diesel tax, pay essentially no tax on diesel fuel, and pay disproportionately lower rates of import duty and annual license fees.

Table 7.9 Road maintenance costs imposed by different vehicle types v. fuel tax

	Road maintenance cost, Rs/ vehicle km	km/ litres	Litres/ km	Fuel tax/km	Ratio
	[1]	[2]	[3]	[4]	=[4]/[1]
Passenger cars(gasoline)	1.35	10	0.10	2.83	210%
Passenger cars(diesel)	1.35	10	0.10	0.29	21%
Vans/4WD/DualPurpose	1.94	8	0.13	0.36	19%
Buses	3.25	3	0.33	0.96	30%
Small trucks	2.66	5	0.20	0.58	22%
Medium trucks(2-axle)	7.06	4	0.25	0.72	10%
Medium truck(3-axle)	11.39	4	0.25	0.72	6%
Articulated truck	8.77	3.5	0.29	0.83	9%

Source: col[1] from Jayaweera (2000), pg.40

In short, there exists a range of arguments for rationalization of fuel taxation, which could be achieved in a number of different ways:[4]

- raising the tax on diesel to the same rate as levied on petrol;
- decreasing the tax on gasoline to the same rate as levied on diesel;
- lowering the tax rate on gasoline and increasing the tax on diesel.

The first option raises significant revenues for the government, but the corresponding increase in diesel fuel would be politically unacceptable. The second option would imply a significant revenue loss for the government, which may therefore also be discarded. However, the third option can easily be structured so that it is revenue-neutral for the government, and therefore merits detailed consideration.

Theory

Consider the demand for gasoline and diesel shown in Figure 7.7. At present, tax is levied at τ_p and τ_d on gasoline and diesel, respectively, with retail prices of P_{op} and P_{od} at quantities Q_{op} and Q_{od}. respectively. If the tax rate is equalised, that is, $\tau_p = \tau_d = \tau^*$ then diesel consumption decreases by ΔQ_d, and gasoline consumption increases by ΔQ_p

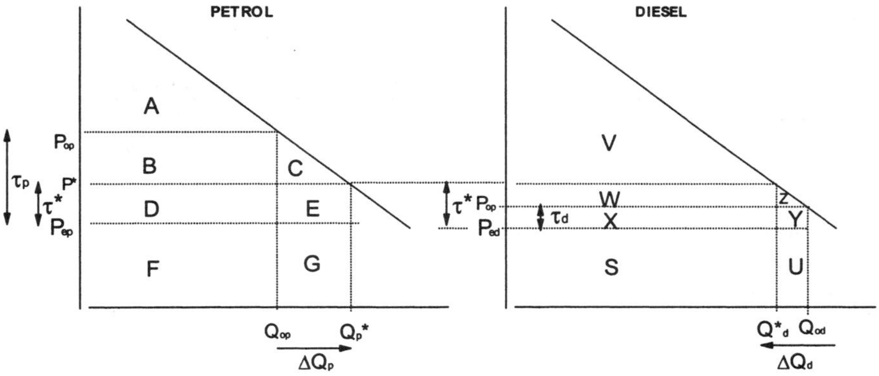

Figure 7.7 Tax equalisation

Under revenue neutrality, the tax revenue before the tax change must equal the revenue with the common tax rate τ^*, and therefore

[B+D]	+	[X+Y]	=	[D + E]	+	[W+X]
old tax on gasoline		old tax on diesel		new tax on gasoline		new tax on diesel
$Q_{op}\ \tau_p$	+	$Q_{od}\ \tau_d$	=	$\tau^*\ Q_p^*$	+	$\tau^*\ Q_d^*$

Hence τ^* follows as

$$\tau^* = [Q_{op}\,\tau_p + Q_{od}\,\tau_d] / [Q_p^* + Q_d^*]$$

where the new quantities Q_p^* and Q_d^* are given application of the price elasticity model discussed in Chapter 5.

However, while the impact on government revenue is (by definition) revenue neutral, the net impact on consumers will be given by the change in consumer surplus (for each user group, which will vary according to the balance of its use of petrol and diesel), and even in aggregate may be non-zero. Table 7.10 enumerates the net economic benefits before and after equalisation. The net (aggregate) impact on consumers calculates to B+C-W-Z, the magnitude of which will be determined by the relative price elasticities involved.

Table 7.10 Net economic benefits of tax equalisation

	Before equalisation	After equalisation	Net change
Gasoline			
Consumer benefit	*A+B+D+F*	*A+B+C+D+E+F+G*	*G+C+E*
-resource cost	*-F*	*-F-G*	*-G*
Net economic benefit	*A+B+D*	*A+B+C+D+E*	*C+E*
Distributed among:			
Consumers	*A*	*A+B+C*	*B+C*
Government [tax]	*B+D*	*D+E*	*E-B*
Diesel			
Consumer benefit	*V+W+X+S +Z+Y+U*	*V+W+X+S*	*-Z-Y-U*
-resource cost	*-S-U*	*-S*	*U*
Net economic benefit	*V+W+X+Z+Y*	*V+W+X*	*-Z-Y*
Distributed among:			
Consumers	*V+W+Z*	*V*	*-W-Z*
Government [tax]	*X+Y*	*W+X*	*W-Y*
Total			
Consumers	*A+V+W+Z*	*A+B+C+V*	*B+C-W-Z*
Government [tax]	*B+D+X+Y*	*D+E+W+Z*	*0*

Box 7.2 Impacts of diesel fuel taxes on food costs

The question of the impact of diesel fuel prices on marketing costs of vegetables was examined by Indraratna (1994), using beans as an example. Though this study analysed the benefit of a diesel price decrease (of March 1994), it found that a one Rs/litre change in diesel price would result in a decrease in delivered cost of vegetables 0.0144-0.024 Rs/kg, compared to a retail price of Rs 25/kg. This was based on the fuel consumption of a round trip from Colombo to Nuwara Eliya (a round trip distance of 360 km, consuming 72 litres) for a saving of Rs 72 per round trip (for a one Rs decrease in price). For 5,000 kg of vegetables (say 100 bags @50kg), this calculates to 1.4 cents/kg (with greater savings at lesser loads). Transportation costs were assessed at 7 per cent of the total marketing cost of beans.

Since 1994, diesel prices have increased from Rs 11.40 to Rs 21.50/litre, and retail bean prices from Rs 25 to Rs 57/kg. In constant 1994 Rs, today's price of beans is Rs 36/kg, and the retail diesel price is Rs 13.4/litre, from which is evident that since the date of the study in 1994, the real price of beans has increased much more than the real price of diesel.

Results

Figure 7.8 shows the results of a tax equalisation on diesel and gasoline under revenue neutrality for the government, expressed as NPVs (at 12 per cent over 25 years), assuming a world oil price of US$ 26/bbl and current retail prices. The resultant rate of (equal) tax is about Rs 7 per litre: it increases slightly from year to year to maintain the net revenue, because the differences in elasticity values for the two fuels causes shifts in the mix.

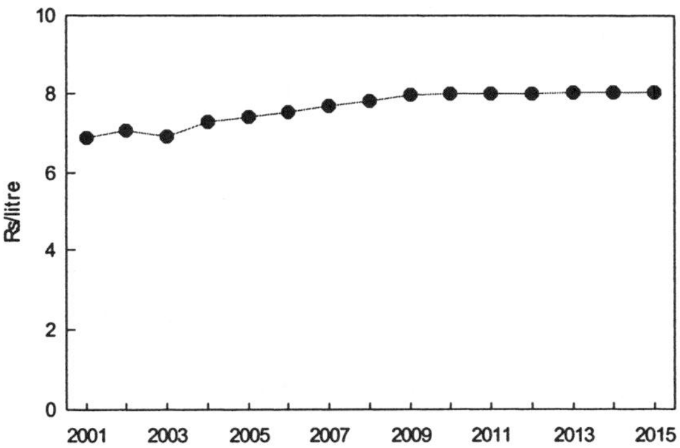

Figure 7.8 Equalised tax rate

Imposition of a diesel tax of Rs 7–8/litre would doubtless be vociferously opposed by the freight industry – which will overlook the fact that, at present, diesel vehicles do not pay an equitable share of road damages, and argue instead that consumer prices would increase – particularly for food. However, as suggested in Box 7.2, this argument would be hard to sustain.

The physical impacts are shown in Table 7.11. There is a net decrease in fuel use, a consequence of the differences in the assumed price and cross-price elasticities. While gasoline fuelled cars and vans are immediate substitutes for their diesel counterparts, it would be much more difficult for lorry owners to shift to gasoline vehicles (since these are largely unavailable in the Indian market that is the principal source), so lorry operators can indeed be expected to utilise their vehicles more efficiently if the price of diesel increases (and hence the net savings in total fuel).[5]

Table 7.11 Environmental impacts of tax equalisation

	Value	2010				Lifetime
	Value $/ton	baseline	+policy	delta	% of baseline	Savings (discounted)
Petrol, 1000 tons		275	304	29	11%	
Diesel, 1000 tons		928	838	-91	-10%	
Emissions						
PM-10, tons	17,790	990	944	-46	-5%	318
SO_2, tons	2,924	9,282	8,377	-905	-10%	6,156
NOx, tons	700	49,278	46,858	-2,420	-5%	16,682
Carbon, 1000 tons	20	1,045	981	-63	-6%	697

Assumptions: world oil price: US$ 26/bbl; price elasticities from Jayaweera (1999).

The distributional impacts of this equalisation are shown in Table 7.12. The net economic benefits of tax equalisation are large. First, there are substantial reductions in road maintenance costs consequent to the reduction in vehicle-km from the heavier diesel vehicles, so the impact on government is positive despite fuel tax revenues remaining unchanged. Second, we see significant environmental benefits, bringing the net economic benefit to Rs 15.8 billion (in NPV terms).[6] In this presentation we make the following assignments among road users:

- low-income = bus and 3-wheeler passengers
- medium-income = motorcycle owners
- high-income = owners of cars and dual purpose vehicles
- other = owners and operators of lorries

It is evident that equalisation of tax rates little affects either low-income or high-income road users: the net impact, expressed as a percentage of their baseline fuel-bill, is less than 4 per cent. In the case of low-income users, bus passengers would suffer the impact of higher diesel fuel prices, but these are also the most typical users of 3-wheelers, who would experience a fuel cost saving of about the same amount.

Thus, the principal beneficiaries are motorcycle users, and the principal losers are lorry owners and land vehicles. In fact, it is clear that the present tax structure is such that the cross-subsidies are not from high-income to low-income road users, as may be supposed, but from motorcycle owners (that is, medium-income) to agriculture and lorry operators. It seems very unlikely that this is the intended effect of the present fuel tax structure. It is also evident that the deadweight losses attributable to the difference in diesel and gasoline price elasticities are in all cases offset by substantial environmental benefits.

Table 7.12 Impact of tax equalisation (Rs million, NPV)

	Road users							Environment		
	Low income (Bus)	Med income (MC)	High income (Car)	Agric.	Freight (Lorry)	Govt	Total	Local	Global	Total
	[1]	[2]	[3]	[4]	[5]	[6]	[7]	[8]	[9]	[10]
Petrol tax	13,410	17,878	14,965	266	2,302	-48,822	0			0
Area[C](petrol)	264	353	291	5	45		958			958
Area[B](petrol)	15,445	20,593	17,233	307	2,652		56,229			56,229
Area[Z](diesel)	-928		-818	-133	-1,153		-3,033			-3,033
Area[W](diesel)	-15,199		-13,496	-2,184	-18,957		-49,836			-49,836
δ [road maintenance]										
Diesel						8,576	8,576			8,576
Gasoline						-3,968	-3,968			-3,968
PM-10							0	770		770
SO_2							0	2,459		2,459
NOx							0	1,591		1,591
Carbon reduction							0		2,073	2,073
Net impact	-418	20,946	3,210	-2,006	-17,414	4,608	8,927	4,820	2,073	15,820
Baseline fuelbill	91,176	43,845	94,484	9,038	73,120					
Net impact as %	-0%	48%	3%	-22%	-24%					

Note: positive numbers=benefits; negative numbers=costs

With regard to the other two options for equalising taxes, if one increases the diesel tax to the same level as that of gasoline, then this would clearly bring larger environmental benefits and greater revenues to the Government, but unacceptable increases in private costs. The alternative of reducing the gasoline tax to that of diesel (that is to essentially zero) is even more infeasible, given the loss of government revenue involved (however popular such a measure might be to road users).

ANNEX:
SUBSIDIES AND DEADWEIGHT LOSSES

Subsidies on petroleum products are widely advocated on grounds that the poor should be protected from bearing the full cost of important items in the household budget. As we have seen in the case of Yemen (Box 7.1), subsidies often do not reach the intended beneficiaries.

Subsidies are what economists term "transfer payments". That is, they represent payments from one group of society to another group – for example in the case of diesel subsidies, from Government (or taxpayers in general) to users of diesel. If diesel subsidies are eliminated, then these same funds will be available to the Government and can be spent by it on cash income support for the poor, or in subsidising other goods, or spent in development projects such as training teachers and health workers and building schools and clinics. In other words, always assuming that the funds would be used for development purposes (and not, for example, for imported military equipment), it is very unlikely that eliminating a subsidy would have significant adverse macroeconomic effects. Indeed, while some jobs may be lost in diesel-intensive sectors of the economy, these will be replaced by new jobs for teachers and nurses, as well as construction jobs resulting from capital spending in health and education, again assuming the incremental Government revenue is wisely spent.

However, regardless of whether a petroleum subsidy reaches the intended beneficiaries, subsidies result in a real cost to society, a cost that economists call "deadweight losses." This occurs because the incremental cost of the subsidy to government exceeds the incremental benefit to the consumer.

The figure demonstrates why this should be so. The downward sloping curve represents the consumer's demand curve: the higher the price, the less he consumes. Now suppose the border price is P, at which price the quantity Q is consumed. The total benefit to the consumer is the area under the demand curve, i.e. the area A + B + C. But the consumer pays P x Q, or the area B + C, so the net benefit to the consumer, called "consumer surplus", is the area A.

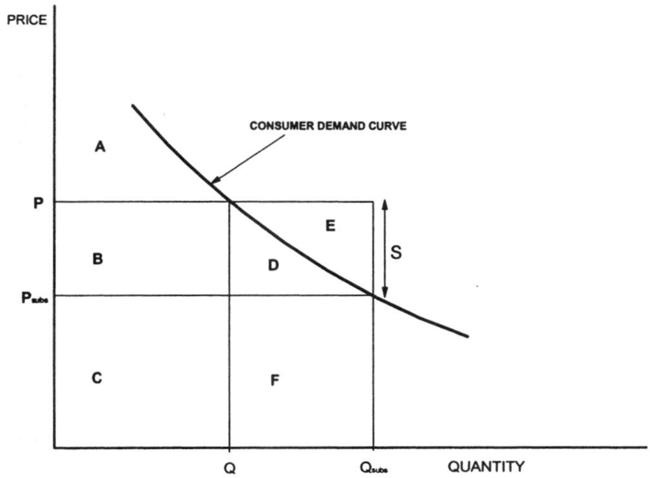

Now suppose the Government decides to subsidize the fuel in the amount S, and sets the domestic retail price at P_{subs}. At this lower price, the consumer increases his consumption to Q_{subs}. Therefore the cost to government is the subsidy times the quantity consumed, S x Q_{subs}, equal to the areas B + D + E.

This cost is offset by an increased benefit to consumers. At the price P_{subs} they consume Q_{subs}, at a cost of P_{subs} x Q_{subs} = C + F. The total area under the demand curve (to Q_{subs}) is A + B + C + D + F, and therefore the net benefit is the difference, namely the area A + B + D. Previously the consumer surplus was A, so the result of the subsidy is an increase in benefit of B + D.

Thus the incremental cost to Government is B + D + E, while the incremental benefit to the consumer is B + D, so the net result is a loss equal to the area E. This cost is the so-called "deadweight" loss, which arises regardless of whether the subsidy reaches its intended beneficiaries.

To calculate the deadweight loss we need an appropriate functional form for the demand curve, for which a simple (and widely used) representation is that of constant elasticity β, namely

$$Q_p = Q_o\left(\frac{P_p}{P_o}\right)^{\beta}$$

The area D+F is then given by the definite integral of this function, namely

$$\frac{P_o}{Q_o^{1/\beta}} \int_{Qo}^{Qp} Q^{1/\beta}\, dQ = \frac{P_o}{Q_o^{1/\beta}} \Big|_{Q_o}^{Qp} \frac{Q^{1/(1+\beta)}}{1+1/\beta}$$

Since the area E + D + F is easily obtained as the product of the level of subsidy S and the incremental consumption Q_{subs}-Q, the deadweight loss E follows immediately as the difference.

NOTES

1 ShellGasLanka estimates the demand (at official filling stations) at about 6,800 metric tonnes, equivalent to about 3 per cent of the gasoline demand (of 220,000 tonnes in 2000). Much of the LPG conversion took place in 1998–1999, during which time LPG demand increased from 99,000 to 140,000 tonnes, as compared to the trend increase of 10–11,000 tonnes per year (which can be attributed to domestic demand). If we subtract 21,000 tonnes from the 41,000 tonne increase in 1998 and 1999, it may be safe to assume that the remaining 20,000 tonnes of LPG increase is attributable to new automobile use, making total use in transportation equal to 27,800 tonnes.

Given 15,500 vehicles converted, the per-vehicle use would be 1.8 tonnes/year, or about 3,275 litres/year. If we take consumption at about 8 km/litre, this implies that these vehicles are driven 26,200 km/year, triple the distance we estimate for gasoline cars on average (8,275 km/year, see Table 2.2). That it is particularly the high-mileage vehicles that are converted LPG is entirely logical, given the high conversion costs. Anecdotal evidence also supports these calculations, for it is extremely rare to encounter a taxi that still runs on gasoline.

2 Colombo is now served by a significant fleet of (yellow) radio-controlled taxis, with a fare of about Rs 30/km, at which price the service is used largely only by the highest income groups (and tourists and businessmen). These taxis have almost all been converted to LPG (of which a significant fraction use domestic bottles).

3 However, hopefully with the completion of the naphtha-fired combustion turbine at Kelanitissa, in future surplus naphtha will be absorbed by the power sector. For a discussion of the economics of naphtha use in power generation, see Meier and Munasinghe (2003a, section 4.1).

4 Others have called for more fundamental changes in the tax structure, not just for auto-fuels, but for all petroleum products and crude oil. The World Bank Petroleum Sector Study (1996a) recommended that the Government introduce a simplified taxation system, replacing the profusion of customs duties, taxes and levies with a single sales tax on the final retail price (p. 21).

This study also proposed (p. 20) that the tax differential between gasoline and diesel be corrected by re-establishing a more appropriate ratio between the two prices so as to reflect the true cost to the economy and suggested a 10 per cent increase in the price of diesel. However, this study did not provide any detailed economic analysis of such a proposal, except to note that such a 10 per cent increase 'would result in a correspondent increase in wholesale prices for basic food commodities in the range of only 0.1 per cent to 0.15 per cent.'

5 The presumption here is that indeed surplus naphtha would have been used by the power sector for power generation, so the economic cost of the additional gasoline imports is taken at the Singapore price plus freight (rather than at Singapore minus freight which would be the relevant cost if naphtha is exported)

6 Note that, unlike in the case of measures discussed elsewhere, fuel savings (or changes in fuel consumption) do not appear in this table (even though there would undoubtedly be lower diesel consumption and somewhat more gasoline consumption). However, since there is an equivalent loss of benefit (area under the demand curve), the two cancel out. However in the case, say, of the expressway project, fuel savings are counted as an economic gain, because these are achieved without diminution of consumer benefits.

8. Introducing unleaded gasoline*

Sri Lanka was long criticised for its gasoline lead content,[1] for which the specification was 0.43 g/litre of added tetra-ethyl lead (TEL). This is significantly above the specification values in other countries, as shown in Table 8.1.

Table 8.1 Comparison of lead in leaded gasoline

	Source	TEL, g/litre
Brunei	ASEAN (2001)	0.11
Indonesia	ASEAN (2001)	0.30
Malaysia	ASEAN (2001)	0.15
Philippines	ASEAN (2001)	0.15
Sinagpore	ASEAN (2001)	0.01
Thailand	ASEAN (2001)	0.01
Vietnam	ASEAN (2001)	0.40
China	Zaidi (2001)	0.35
Pakistan	Yangsheng (2001)	0.35
Sri Lanka		0.43

Yet the addition of TEL, as may readily be calculated from refinery data, has long been significantly lower (Table 8.2), so in fact the difference to other Asian countries has been less than suggested.[2] Moreover, some of the earlier estimates of lead emissions are also in demonstrable error due to poor assumptions about the vehicle fleet, fuel efficiency and annual vehicle-km.[3]

Table 8.2 Actual average TEL additions (based on refinery data)

	1993	1994	1995	1996	1997	1998	1999	2000	2001
g/litre	0.17	0.25	0.23	0.30	0.31	0.32	0.25	0.15	0.11

Thus our estimates of lead emissions, even if unleaded gasoline were not introduced, is substantially lower than those estimates by the Clean Air 2000 study (Figure 8.1).

* Chapter 8 contributor: A. Abeywardena.

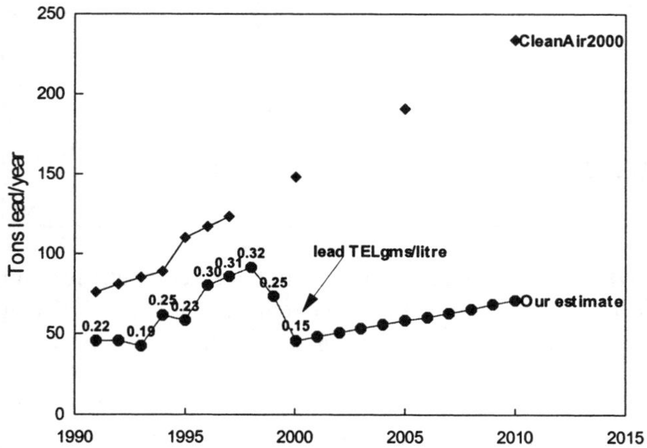

Source: Clean Air 2000 Action Plan (MEIP, 1994); our estimates based on CPC historical data plus forecast of vehicle fleet based on estimates of Table 2.3.

Figure 8.1 Estimates of lead emissions from automobiles

Unleaded gasoline was first introduced in June 1997, with the hope that by 2000, unleaded might account for 30 per cent of sales.[4] But sales have been disappointing, as shown in Table 8.3; in 2000, less than 2 per cent of total gasoline sales were unleaded. Several explanations for low sales are likely:

- few outlets sell unleaded gasoline (only 12 of 639 outlets in Sri Lanka provide unleaded petrol, mainly in Colombo);
- price differential between leaded and unleaded (unleaded sells at Rs 53/litre as opposed to Rs 50/litre for leaded);[5]
- conversion of vehicles that need 95 octane unleaded to LPG (whose octane is also about 95).

Table 8.3 Sales of unleaded gasoline

	1997	1998	1999	2000
Total petrol (tonnes)	194,828	204,351	213,068	224,379
Unleaded gasoline (tonnes)	1,582	3,068	3,476	4,160
Unleaded gasoline (%)	0.8	1.5	1.6	1.9
Average TEL, g/litre	0.314	0.318	0.25	0.148

CPC has argued that to the extent that unleaded gasoline involves higher production costs (of about Rs 1.25), the higher sales price is justified. Moreover, to add unleaded gasoline at an outlet costs about Rs

475,000 (tank, dispensing machine, construction) per station.[6] There are also additional tank-farm costs at the storage facility end.

However, it must be understood that the introduction of unleaded gasoline by CPC was in response to a perceived customer demand for high-octane (95) unleaded petrol, rather than to environmental concerns per se. This demand was from high-end (luxury) vehicle[7] owners, whose ability and willingness to pay for the incremental cost is clear. Therefore, to the extent that customer demand was the rationale, the price premium for unleaded is not unreasonable, given its higher octane.

8.1 THE MYTHS

Most of the traditional objections to the introduction of unleaded gasoline have been heard in Sri Lanka:

- that, as a poor developing country, the costs are too high;
- that it has a large fleet of older vehicles designed for leaded gasoline which depend upon lead as a lubricant (to avoid so-called 'valve seat recession' in older cars that have soft valves and valve seats);[8]
- that if unleaded gasoline is produced by running reformers at higher severity, it would have a higher benzene content (a known carcinogen).[9]

These arguments have little merit.[10] The consequences to health of lead in gasoline have been demonstrated beyond reasonable doubt, and arguments to the contrary (for example, that the observed decline of lead concentrations in blood are attributable to declining use of other lead compounds, such as lead-based paints, rather than to the introduction of unleaded gasoline) have been convincingly rejected.[11] We have already noted that a conservative order of magnitude estimate for the health damage for lead in gasoline is around Rs 7/litre gasoline, which is a small fraction of the estimated incremental cost of around Rs 1–3/litre (as demonstrated below). Finally, many international sources also argue that even aside from the health benefits, the incremental costs of unleaded gasoline are largely offset by reduced maintenance costs.

With respect to valve seat recession, there is broad consensus in the literature that this is a greatly exaggerated problem, and to the extent it has any validity at all, applies primarily to the former Soviet Union and Eastern European countries where car manufacturers did not introduce hardened valve seats until very recently.[12] For the past twenty years, the overwhelming majority of gasoline vehicles in Sri Lanka have been imported (as 2–3-year-old reconditioned vehicles) from Japan, where leaded gasoline was phased out in 1980, and where all models have long had hardened valve seats. Indeed, upon import, their catalytic converters are removed in order to use leaded gasoline (the avoided costs of which constitute a further economic benefit of unleaded gasoline). Thus all domestic Japanese engines were designed to use unleaded gasoline from 1975 onwards, and these constitute the bulk of the present gasoline car vehicle stock in Sri Lanka.[13] Moreover, in a number of countries, leaded gasoline has been phased out despite a large number of vehicles with soft valve seats: for example, in the Slovak Republic, 70 per cent of the fleet had soft seals at the time of conversion, a problem solved by use of a special purpose additive.

With regard to the benzene issue, two points are worth noting.[14] The first is that the health impacts of benzene exposure are far smaller than those of lead exposure: the United States Environment Protection Agency (USEPA) estimated that, for the USA, 47 cases of leukaemia deaths could be attributed to benzene in gasoline, as opposed to 4 000–5000 deaths avoided by the reduction of lead in gasoline (USEPA, 1985). Second, the most significant share of gasoline-related exposures to benzene is related to evaporation in garages, during driving, and at filling stations; and that, in any event, the main source of exposure to benzene in the population is from smoking.

8.2 TECHNICAL ASPECTS

Figure 8.2 illustrates the relevant parts of the refinery involved in production of gasoline.[15]

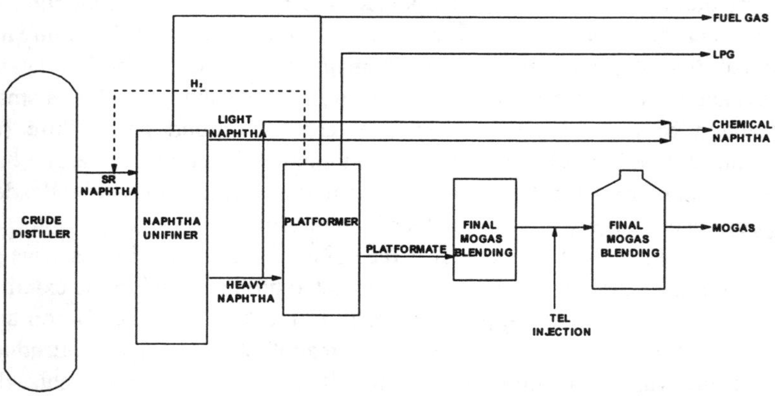

Figure 8.2 Refinery configuration

Figure 8.3 shows the effect of TEL addition to different blendstocks on Research Octane Number (RON). Note the non-linear response: the first 0.1g/litre typically raises the RON by a much larger amount than from 0.1 to 0.2g/litre. Studies show that the incremental difference between leaded and unleaded gasoline (at the same RON) is much smaller than the difference between octane levels. At 90 RON the incremental cost of unleaded gasoline is only Rs 0.08/litre (see Table 8.4), while at 95 RON the incremental cost of unleaded is higher (Rs 0.14-Rs 0.22/litre).[16] However, the difference between 90 RON and 95 RON is between US$ 1.20 to US$ 1.40/bbl (about Rs 0.67 – Rs 0.80/litre).

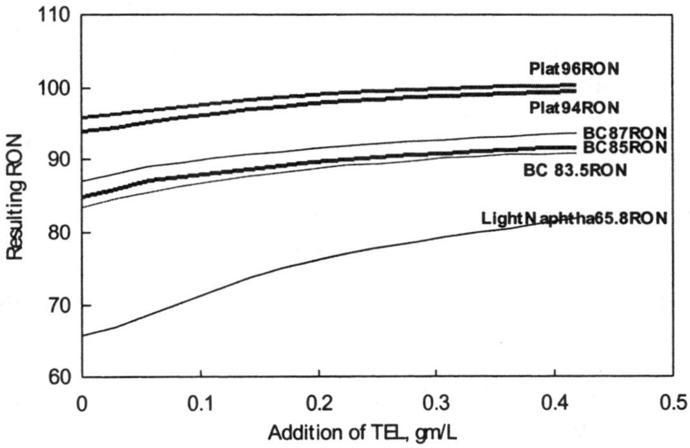

Figure 8.3 Effect of TEL additions

Previously, gasoline was blended from two stocks: platformate, typically of 92 octane, and light, straight-run naphtha, typically of 67 octane, blended to produce gasoline of around 84 octane, and brought to 90 octane by the addition of around 0.25 g/litre TEL.

The higher cost to CPC of making unleaded gasoline derives in part from the need to run the platformer at higher severity (though a by-product is more hydrogen, which has its uses for gasoil desulphurisation). Also, since smaller amounts of light naphtha are required for blending, it increases the amount of the surplus that must be exported (at a significantly lower price than gasoline) – though that is no longer true once the naphtha can be used for power generation. Finally, the volume of unleaded gasoline that can so be produced is also reduced, which may necessitate importing gasoline (at the same time as naphtha is exported).[17]

Table 8.4 shows the results of the optimisation analysis of the costs of producing unleaded and leaded gasoline at various Octane numbers and under a platformer constraint.[18] This analysis uses the TEL blending curves of Figure 8.3, and uses average 2000 border prices for the inputs.

Table 8.4 Incremental costs of unleaded gasoline

		90RON	95RON	95RON (with platformer constraint)(a)
Cost of unleaded gasoline	(US$/bbl)	29.72	30.96	31.16
Delta (octane improvement)	(US$/bbl)		1.24	1.44
Cost of leaded gasoline	(US$/bbl)	29.59	30.77	30.77
Delta (octane improvement)	(US$/bbl)		1.18	1.18
Incremental cost	(US$/bbl)	0.13	0.26	0.39
	(Rs/$bbl)	12.06	23.4	35.1
	(Rs/litre)	0.08	.14	.22
	(US cents/litre)	0.085	.16	.25

Note: (a) Platformer severity limited to maximum 96 octane

The incremental production costs of 95 RON over 90 RON at Sapugaskanda are also consistent with the price premiums observed on Singapore spot markets, shown in Figure 8.4. For most of 1997–99, the cost of moving from 92 RON to 95 RON was around US$1/bbl (or Rs 0.6/litre at Rs 90/US$) – which has more than doubled since then. And 97 octane commands even higher premia.

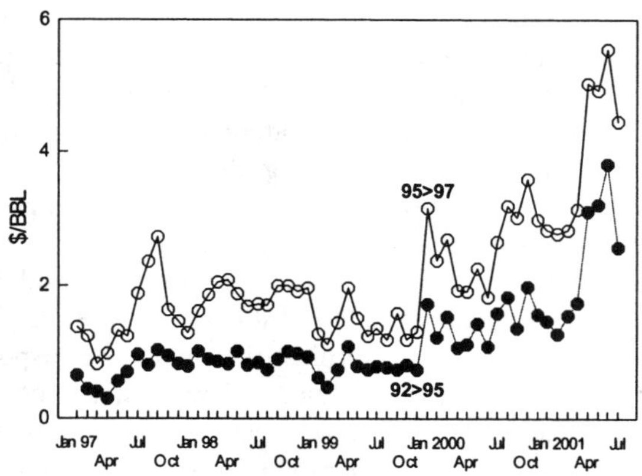

Figure 8.4 Octane premia (Singapore spot market prices)

8.3 TAX INCENTIVES

The efficacy of an appropriate taxation scheme for the introduction of unleaded gasoline is evident from the experience of Hong Kong. According to Walsh (1992), when Hong Kong introduced unleaded gasoline in Spring 1991, prices were modified to make unleaded gasoline cheaper to the consumer than leaded. As a result, unleaded sales soared from zero to over 50 per cent within the first month. If indeed unleaded petrol is priced less than leaded petrol, then no rational consumer would use the latter, and unleaded gasoline consumption would be constrained only by the ability of CPC to supply adequate unleaded gasoline, and, perhaps more importantly, by the number of stations offering unleaded.

The relationship between price differentials and market share is well established statistically. A World Bank study (Lovei, 1998) showed the relationship between market share for unleaded (Y), price differential (P), and share of the fleet with soft valve seats (V) as:

$$Y=61.7 + 3.41\,P - 1.19\,V;\ R^2=0.78$$

Thus a 1 percentage increase in the price differential resulted in a 3.5 per cent increase in the market share for unleaded.

At this point the CPC is committed to phasing out lead in gasoline. The real question, therefore, is not so much whether this should or should not be done, but how it should be done. We know of no examples in Asia of where unleaded gasoline was successfully introduced while selling at a higher price than leaded. For example, in Thailand, leaded gasoline was phased out over a 4-year period (1991–95), with differential tax rates used to secure the desired result (Figure 8.5).[19] Although the ex-refinery price of unleaded is higher (by about 15 per cent), the retail price of unleaded is lower by about 2.5 per cent.[20]

Several aspects of the Thai experience seem worth noting. First, the differential between unleaded and leaded gasoline does not have to be large: even a small price advantage attributed to unleaded (of 2.5 per cent) is sufficient to induce motorists to choose unleaded – always assuming there are a sufficient number of retail outlets to supply it. Second, the volatility in the world oil price far exceeds the ex-refinery production cost differential of 15 per cent equal to about 2 US cents/litre.[21] In 1995, the average Singapore spot price for gasoline was around US$ 23.5/bbl, or 14.8 US cents/litre; it then fell to US$ 13/bbl (8 cents/litre) in December 1998, and increasing again to US$ 38/bbl (US 24 cents/bbl) in August 2000. Therefore, even if the entire price differential were passed onto the consumer, this would be a modest increase compared to any adjustments required by cif prices.

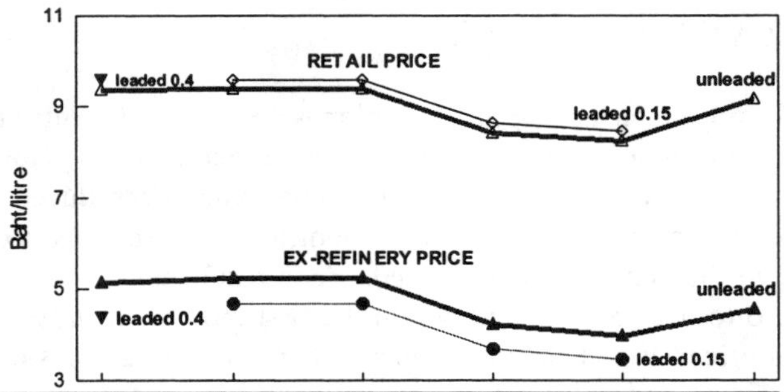

	1991	1992	1993	1993	1994	1995
▼ leaded 0.4	4.4					
● leaded 0.15		4.7	4.67	3.7	3.5	
▲ unleaded	5.1	5.3	5.25	4.2	4.0	4.6
▼ leaded 0.4	9.59					
⊖ leaded 0.15		9.59	9.59	8.62	8.44	
▲ unleaded	9.38	9.41	9.41	8.43	8.24	9.18

Source: Vivekaphirat (2001), p. 18

Figure 8.5 Implementing unleaded gasoline in Thailand [Baht/litre]

8.4 ECONOMIC ANALYSIS

Table 8.5 shows the results of the analysis, under the following assumptions:

- An incremental production cost at the refinery of Rs 0.8/litre (assuming the platformer is limited to maximum severity of 96 octane).
- Lead health damage equivalent to Rs 7/litre.
- Infrastructure development costs based on a 3-year programme to extend service of two grades of gasoline to an additional 400 filling stations.[22] This adds about Rs 0.25/litre to the cost.

It is evident that the incremental environmental benefits of unleaded gasoline are large compared to the incremental costs. Moreover, the increased costs experienced by consumers (assuming the incremental costs are passed onto them) are small, less than 2 per cent of the present fuel bill: but when the health benefits are apportioned back to road users, benefits particularly to motorcycle riders are large. In addition, the occupational groups most affected by present lead exposure – trishaw drivers, traffic wardens, and policemen – would be directly benefited.

Table 8.5 Impact of unleaded gasoline

	Road users								Environment		
	Low income	Med income	High income	Agric	Business	CPC	Govt	Total	Local	Global	Total
Economic costs								0			0
Area [C] (petrol)	-37	-49	-42	-1	-7			-135			-135
Price increase	-756	-1,048	-698	-12	-109	2,623		-0			-0
Refinery costs [B]						-2,623		-2,623			-2,623
Infrastructure costs						-298		-298			-298
Tax impacts							-241	-241			-241
Environment											
Lead								0	19535		19,535
PM-10								0	10		10
SO_2								0	-13		-13
NOx								0	17		17
Carbon reduction								0		14	14
Net impact	-792	-1,098	-739	-13	-115	-298	-241	-3,297	19,549	14	16,266
Apportioned health impact											
Diesel											
Petrol	5,632	7,815	5,199	91	811						
Net impact	4,840	6,717	4,460	78	696						
Baseline fuel bill	122,900	50,998	127,271	12,947	105,586						
Net impact as a percentage	-1	-2	-1	-0	-0						
Percentage (after health)	4	13	4	1	1						

Thus it is hard to avoid the conclusion that the introduction of unleaded gasoline is highly cost effective. The switching value for the damage costs is around Rs 1/litre, which is a small fraction of our already very conservative estimate of Rs 7/litre (in turn only 5 per cent of the value based on US studies adjusted for population and GDP).

Moreover, we show very small incremental refining costs. Experience in other countries shows little problem with old cars (that are said to require lead for lubrication), so the costs of duplicating infrastructure could be eliminated by simply phasing out leaded gasoline straight away. Since most gasoline vehicles in Sri Lanka are of Japanese domestic origin, where lead was phased out 20 years ago and vehicles have had hard valve seats for the same period, the number of vehicles that actually require such lubrication is very small. These vehicles could be supplied with a special purpose additive at a minute fraction of the cost of duplicating infrastructure (following the successful experience in the Slovak Republic). Of course, a modern supply system would need to offer two grades of gasoline (say 92 and 95 octane unleaded). But to charge these incremental infrastructure and octane costs to the environmental account is hardly reasonable.

NOTES

1 Specifications for lead in many African countries are much higher than in Asia: for example, in Benin, Cameroon, Chad, Uganda, Zimbabwe, the (1996) value is 0.84 g/litre. For a complete worldwide list, see, for example, M. Lovei (1998), Annex A.

2 The World Bank Transport Sector Study (World Bank, 1996b) is typical: 'Leaded gasoline is also used: about 0.43 g/litre is allowed which is high in comparison to leaded gasoline in other countries.' There is no mention of the actual rates of TEL addition. Even the 1994 report of the Sub-committee on Fuel Reformulation, charged with an examination of fuel formulation issues raised by the Clean Air 2000 programme, makes the same statement: ' . . . at present we use about 0.45 g lead per litre of petrol. This is about 3 times higher when compared to the leaded petrol in other countries.'

3 See earlier text for a discussion of the problems in the USAID studies of the early 1990s that were the basis for the Clean Air 2000 programme.

4 See Report of the Subcommittee on Fuel Reformulation, 1994: para 6.1.

5 Initially, the price differential was Rs 5/litre, which was subsequently reduced to Rs 3/litre.

6 This estimate, say 500,000 x 300 (say 50 per cent of all outlets) = Rs 150 million is considerably higher than that assumed by the Sub-committee on Fuel Reformulation (Rs 35 million).

7 Typically of European manufacture, for example, Volvo, Mercedes.

8 CPC, Environmental concerns as a result of increase in use of petroleum fuels, May 1992: states that 'The old cars in use in Sri Lanka will need minimum lead level of 0.15

g/litre to ensure valve lubrication.' In fact, a number of studies show that minimum levels are far less than this figure, around 0.05-0.07 g/litre.

9 Sub-committee on Fuel Reformulation, Government of Sri Lanka, op.cit., 'use of unleaded petrol without a catalytic converter will increase the aromatic organic compounds coming to the atmosphere from petrol'.

10 Similar myths have arisen in most developing countries. For example, in the context of the Philippines they are rehearsed, and answered, at the website www.motorista.com.ph/motortips/102099html. See also Shell-Philippines website www.shell.com/phen/content 0,4600,31743-57201,00.html .

11 See, for example, the literature review by Thomas 1995).

12 See, for example, Lovei (1998), and UNEP (1998). As one might expect, many of these arguments have been advanced by the world's leading supplier of lead additives.

13 Finally, as noted by Shell-Philippines, op.cit., most cars designed for leaded gasoline have 'lead memory' that typically affords protection for a further 10,000 to 20,000 km after a switch to unleaded gasoline. The very small percentage of older soft valve cars that remain in Sri Lanka (such as vintage Morris Minors) are unlikely to be driven very much, and are consequently unlikely to experience significant damage. These few vehicles could easily use some other suitable additive to provide the additional lubrication.

14 For a more detailed discussion see, for example, Lovei (1998): Box 4, 'The benzene myth'.

15 Light naphtha is treated in a "Unifiner", equivalent to hydrodesulphurisation, where hydrogen gas is used to react with hydrocarbons, breaking off sulphur as hydrogen sulphide and lesser amounts of other sulphide compounds such as methyl sulphide.

16 This is significantly lower than earlier estimates by CPC of additional costs of Rs 1.25/litre. However, the basis for those earlier estimates is unclear.

17 Obviously this raises broader questions of the profitability of refining at Sapugaskanda, and whether all products should simply be imported in whatever proportions required by the domestic market. However, these questions lie beyond the scope of this book.

18 See Meier and Munasinghe (2003b), Section 9, for details.

19 Vivekaphirat (2001). In addition to normal taxes, the Thai Oil Fund was used as a cross-subsidy measure in this tax equalisation.

20 The impact on ambient lead levels was dramatic, falling from about 0.5 $\mu g/m^3$ in 1990 to less than 0.05 $\mu g/m^3$ by 1997; and roadside levels falling from 2 $\mu g/m^3$ to about the same ambient value of 0.05 $\mu g/m^3$ (Vivekaphirat, 2001, p. 27).

21 In the 1990–96 time period (before the crisis in 1997), the exchange rate was around 25.3 Baht/US$.

22 Assuming Rs 500,000 per filling station plus another Rs 500,000 for each filling station as CPC's incremental development cost (for storage and handling).

9. Options for GHG emission reductions*

This chapter presents the options for GHG emission reductions. We begin with the power sector, summarizing the detailed report on the Sri Lanka Global Overlay (Meier and Munasinghe, 2003c). We then assess transportation sector options are then assessed.

9.1 THE POWER SECTOR

The GHG Overlay Guidelines are taken as the general analytical framework for the analysis of GHG mitigation options (World Bank, 1997d). Among the GHG abatement methodologies presented in the Guidelines, the proposed approach falls into the category of 'bottom-up' studies. The guidelines suggest the following steps:

1. estimate GHG emissions for the reference (business-as-usual) scenario;
2. estimate GHG emissions for a reform scenario to assess the impact of pricing, policy and other interventions typically undertaken with World Bank assistance;
3. identify, evaluate and screen a range of GHG mitigation options;
4. formulate alternative mitigation scenarios and estimate the GHG impacts of each scenario, and compare to the reform scenario to gauge the cost-effectiveness of the GHG emissions reduction options.

In addition, the Sri Lanka Overlay Study developed an analytical framework to deal with significant uncertainties facing Sri Lankan decision-makers in responding to the opportunities offered by the Kyoto Protocol's flexibility instruments (that is, Clean Development Mechanism, Joint Implementation, and eventually emissions trading). This is presented in Chapter 10.

Step one begins with an estimate of GHG emissions from the power sector under a reference (or business-as-usual) scenario. The least-cost capacity expansion programme is developed to satisfy the growth of

* Chapter 9 contributors: Shavi Fernando and J. Nanthakumar.

electricity demand for the next twenty years, while meeting acceptable levels of reliability and environmental protection standards. Next, the GHG emissions associated with a reform scenario (including various policy options) are estimated. Several mitigation options are screened and then combined within alternative GHG mitigation scenarios, which are compared to the reference and reform scenarios. The modelling was done in ENVIROPLAN,[1] developed in the early 1990s for use in a study of environmental issues in the power sector in Sri Lanka (Meier and Munasinghe, 1994), but which has since seen application in many other countries.[2]

Baseline Emissions

Even today, biomass accounts for the largest share of primary energy (Table 9.1). However, the trend towards a greater share of commercial energy is persistent and unmistakable. Since 1984, commercial energy has grown at an annual rate of 4.7 per cent, while over the same period biomass energy has declined at an annual rate of 0.6 per cent.

Table 9.1 Shares of total primary energy

	1984		1996		Change
	1,000 TOE	%	1,000 TOE	%	84-96 %/year
Biomass	4,203	71	3,925	57	-0.6
Commercial	1,700	29	2,952	43	4.71
Total	5,903	100	6,877	100	1.28

Note: TOE = tonnes of oil equivalent. 'Commercial' includes hydro and petroleum. The data need to be interpreted with care, because 1996 was a drought year, in which the share of hydro was lower than normal. However, 1996 is the latest year for which a complete energy balance is available.

Source: Energy Conservation Fund (1996).

These circumstances are reflected in the results of a comprehensive GHG emissions inventory prepared by the Ministry of Forest and Environment (Ratnasiri, 1998), shown in Table 9.2. However, the reported increases in power sector emissions from 1990 to 1992 are due solely to the variation of hydro generation (1991 and 1992 were dry years), rather that a change in generation capacity (since there were no capacity additions in 1990–92).

Table 9.2 Estimates of CO_2 emissions by source, 1990–92 (000 tonne)

	1990	1991	1992
Power (CEB)	8.5	226.6	643.2
Transport	2,214	2,317	2,334
Energy generation in industry	559	485	561
Commercial (fisheries, agriculture)	172	161	166
Domestic (LPG, kerosene)	550	578	651
Cement/lime production	249	244	436
Refinery own use	0.7	0.6	0.7
Total energy and industrial processes	*3,899*	*4,120*	*4,894*
Biomass	21,261	20,095	21,498
Emissions from agriculture, land-use and waste	8,435	8,370	8,098

Source: Ratnasiri (1998)

This situation will almost certainly change. As noted previously, the share of hydro-electricity in Sri Lanka's energy mix will decline, as almost all new power plants presently anticipated as being necessary over the next two to three decades will be fossil fuelled. Wind power, mini-hydro and DSM will also increase in importance in the near future, but a substantial increase in the use of fossil fuel for power generation seems inescapable. Consequently, in the business-as-usual case, we project that the share of GHG emissions accounted for by the power sector will increase from 18 per cent of energy sector emissions in 1998 to 45 per cent in 2018; in absolute terms the power sector will account for 20.1 million tonnes per year (mtpy) by 2018, compared to 1.7 mtpy in 1997, a more than tenfold increase.

Emission Coefficients

Wherever possible, emission coefficients are calculated by stochiometry, since this is the best way to ensure consistency between assumptions about fuel characteristics, generating plant efficiency, and assumptions about combustion technology and the pollution control options as may be applied.[3] Where such stochiometric calculations are not possible, we use the emission coefficients provided by IPCC. Aggregations to carbon equivalent are made on the basis of the IPCC guidelines shown on Table 9.3. In this chapter we use the standard 100-year equivalents.[4]

Table 9.3 Direct global warming potentials (GWP) of gases

Time horizon	20 years	100 years	500 years
Carbon dioxide	1	1	1
Methane	62	24.5	7.5
Nitrous oxide	290	320	180

Source: World Bank (1997d)

Table 9.4 describes the assumptions and coefficients used for each air emission.

Table 9.4 Emission calculations

CO_2	Stochiometric calculation based upon the carbon content of the fuel. If C is the weight percentage of carbon in the fuel and f is the fraction of carbon oxidized in combustion, then the kg of C emitted per kg of fuel consumed = C f.
SO_2	Stochiometric calculation based on sulphur content in fuel. If S is the weight percentage of sulphur in the fuel; λ is the fraction of SO_2 in the flue gas removed by FGD; and f is the fraction of sulphur oxidized in combustion, then the kg of SO_2 emitted per kg of fuel consumed = $S\,f\,(1-\lambda)\,1.998$.
Particulates	Stochiometric calculation based upon the ash content of the fuel. If A is the weight percentage of ash in the fuel; f is the fraction of ash to fly ash (and 1-f the bottom ash fraction); and γ is the fraction of particulates removed by the ESP, then the particulate emissions per kg of fuel consumed = $A\,f\,(1-\gamma)$.
NOx	Emission factor as provided by the equipment manufacturer (and/or) as per the project specific Environmental Impact Assessment.
CH_4	Emission factor as per IPCC guidelines.

Note: (a) The fraction of carbon oxidized is as given in the Overlay Guidelines (p. 21): 0.98 for coal, 0.995 for gas, 0.99 for oil combustion.

The Business-As-Usual Scenario

The Ceylon Electricity Board (CEB) conducts an annual long-term generation expansion planning study that assesses investment requirements in the power sector as a function of a range of input assumptions on load growth, fuel prices and technology assumptions (CEB, 1998b, 1999). The 'base case' as presented in the 1999 CEB study is adopted for the GHG overlay as the business-as-usual case.

As discussed in Chapter 2 (recall Figure 2.9) electricity demand growth and GDP growth are highly correlated. Econometric studies conducted by CEB for the 1998 CEB demand forecast shows GDP growth to be the most significant determinant of demand (CEB, 1998). Table 9.5 shows the GDP assumptions for the forecast.[5]

Table 9.5 GDP growth assumptions for the CEB forecast

	Forecast rate of annual GDP growth %
1995 (actual)	5.5
1996 (actual)	3.8
1997 (est.)	6.0
1998	6.5
1999	7.0
2000	7.5
2001	8.0

Source: Public Investment Programme 1997-2001: Department of National Planning

In the business-as-usual demand forecast, the tariff is assumed to increase with inflation (no increase in real terms). By 2018 the busbar peak demand is estimated to grow to 4,292 MW (compared to 1,136 MW in 1998); the corresponding 2018 busbar energy demand is 20,678 GWh, up from 5,683 GWh in 1998. Total T&D loss rates, and the system load factor (SLF), are taken as 17.6 per cent (of generation) and 55 per cent, respectively, which are taken at their 10-year average values (both have stayed remarkably constant over the past decade). Of the total T&D loss, 15.3 per cent is assumed to be technical loss, and the balance of 2.3 per cent is assumed to be non-technical (commercial) loss.[6] Figure 9.1 shows the resulting capacity expansion plan for the business-as-usual case.

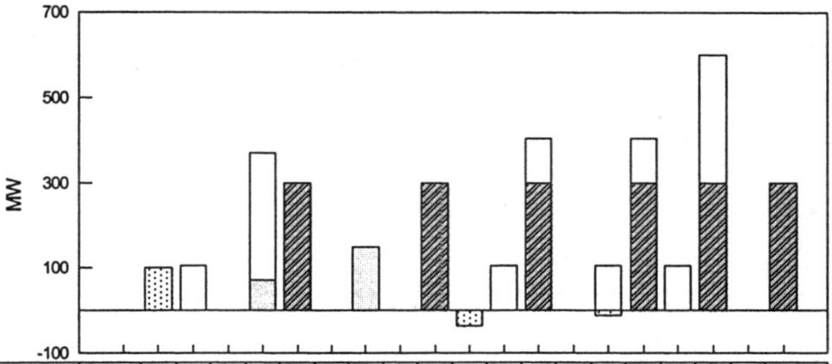

	1999		2001		2003		2005		2007		2009		2011		2013		2015		2017	
hydro					70			150												
coal						300				300			300			300		300		300
liquid fuels			106		300							105	105		105	105	105	300		
gas																				
diesel		100									-36				-13					
renewables																				

Note: Negative figures indicate retirement

Figure 9.1 Capacity expansion plan

The Reform Scenario

In Sri Lanka, the specific components of a comprehensive reform initiative have yet to be articulated – unlike the Indian states of Haryana, Orissa and Andhra Pradesh, where the packages for policy reform measures underway with World Bank assistance have already been identified. Consequently the individual programmes that would be part of a reform and restructuring effort in Sri Lanka will themselves need some screening. In many cases we cannot say *a priori* whether a particular measure (for example, the development of grid connected mini-hydro), would be part of a reform programme based upon purely the economic returns, or whether its economic returns are below the hurdle rate and would be justified only as part of GHG mitigation efforts. Nevertheless, it seems reasonable to suppose that the following elements might be part of the power sector reform programme:

1. A more aggressive set of assumptions for T&D loss reduction (as might be the consequence of distribution privatisation);
2. Implementation of the remaining cost-effective DSM options;
3. Improvements to time-of-day tariff;
4. An appropriate set of tariff reforms (that would set the tariff based upon a return on equity criterion higher than the present criterion for the government-owned Ceylon Electricity Board).[7]

Table 9.6 shows the assumptions for the reduction in technical and non-technical losses in the reform scenario. In light of LECO's reduction in T&D loss rates achieved over the past decade (see Box 1.1), the 12.5 per cent target should also be achievable for Sri Lanka as a whole.[8]

Table 9.6 Assumptions for T&D losses: reform scenario

	Technical	Non-technical	Total
1999	15.5	2.3	17.8
2000	15.3	2.3	17.6
2001	15.3	2.3	17.6
2002	15.0	2.1	17.1
2003	14.0	2.1	16.1
2004	13.0	1.9	14.9
2005	12.0	1.7	13.7
2006	11.5	1.5	13.0
2007 and beyond	11.0	1.5	12.5

With regard to DSM we assume that in the first stage of reform, DSM would be limited to lighting. A recent (and comprehensive) analysis of DSM in Sri Lanka is that of Shrestha et al. (1998). As indicated in Table 9.7, all the appliances examined are economic (as measured by the cost of conserved energy being less than the long-run marginal cost – LRMC), but only residential lighting has a net annual benefit to the utility. This is because the tariff in commercial and industrial sectors is higher than the long-run marginal cost or LRMC (and therefore results in disproportionate revenue losses in consequence of energy savings).[9]

For the reform case DSM programme all of the lighting measures are included: energy efficiency motors and efficient AC systems are considered later as potential mitigation options. By 2007, lighting DSM reduces the peak evening demand by 330 MW, about 15 per cent of the total evening peak load.

As shown in Figure 9.2, the combination of reduced T&D losses and the peak-shaving impact of the lighting DSM programme has a significant impact on the capacity expansion plan. Over the planning period there is a reduction of the number of required baseload coal units from six to five, and a significant reduction in the number of thermal peaking units required towards the end of the planning horizon (needing two rather than six 105 MW OCCTs from 2010 to 2018). This reduction in thermal peaking would be expected given the effect of lighting DSM on the evening lighting peak load.

Table 9.7 Economic and utility costs of efficient appliances

		Economic cost US cents/kWh	Utility net annual benefit US$ million
Residential	13 W CFL	4.21	1.3
	18 W CFL	3.02	2.0
	23 W CFL	1.99	3.7
Commercial	36 W slim tube	0.67	-0.1
	18 W CFL	2.48	-0.7
	Efficient AC	6.21	-2.9
Industrial	36 W slim tube	0.67	-0.3
	18 W CFL	2.48	-3.3
	Energy efficient motors	0.89	-35.6

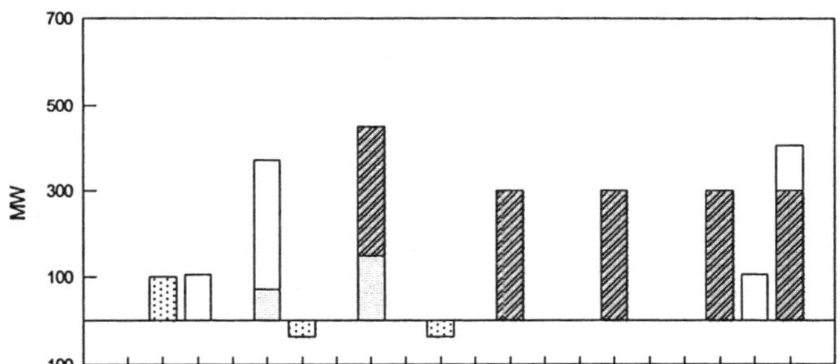

	1999		2001		2003		2005		2007		2009		2011		2013		2015		2017	
hydro					70			150												
coal								300				300			300			300		300
liquid fuels			106		300														105	105
gas																				
diesel		100				-36				-36										
renewables																				

Source: Shrestha et al., 1998, Tables 3 and 4

Figure 9.2 Capacity expansion plan: reform

These adjustments result in a significant reduction in GHG emissions, from 118 million tonnes over the planning horizon in the base case to 98 million tonnes under reform (Figure 9.3).

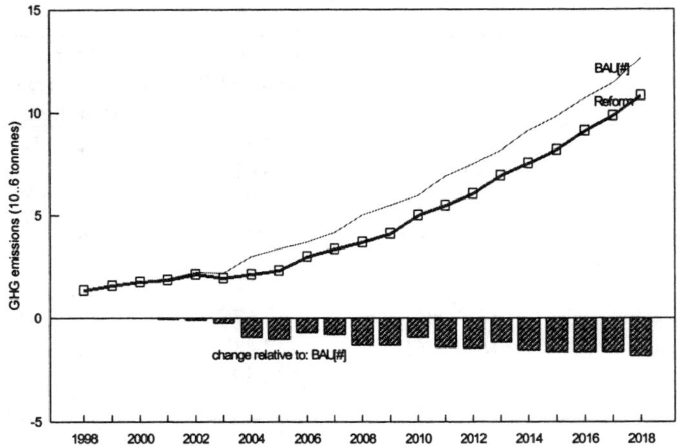

Figure 9.3 GHG emission reductions

The impact of reforms on system costs and GHG emissions can be displayed as a trade-off plot as indicated in Figure 9.4.[10] The impact of T&D loss reduction and DSM is shown separately (each as perturbations of the BAU case). Reform is clearly 'win–win'; indeed, reform is also win–win with respect to other environmental attributes for local air emissions.[11]

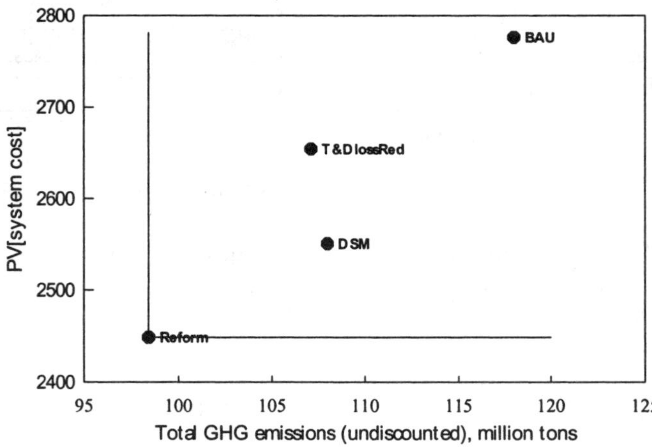

Figure 9.4 Impact of reform

The economic and environmental benefits of power sector reform in South Asia have been confirmed in several Indian studies as well (Box 9.1). Even though air pollution emissions may increase in absolute quantity, their location shifts from self-generation in relatively wealthy populated urban areas to utility-based plants that tend to be in more remote rural areas – for example, in the case of Karnataka, the location of emissions shifts from big cities such as Bangalore to sparsely populated Northern Karnataka or the west coast where larger coal units would be sited.

Box 9.1 Power sector reform in Karnataka

A recent study of Karnataka suggests similar economic and environmental benefits of power sector reform (World Bank, 2004b). Because the Karnataka reform programme has been in place for several years, the Karnataka study compared a business-as-usual scenario (that is, continuing the present path of reform) with a 'stalled reform' scenario, in which the reform steps were frozen at their 2002 values (that is, no further tariff increases in agriculture, no further reduction in technical and non-technical losses, lack of creditworthiness prevents further IPPs from attaining financial closure).

Using a similar methodology as used for the Sri Lanka GHG Overlay Guidelines, the economic costs of stalled reform, as NPV, were estimated at Rs 8,900 crores (about US$2 billion), plus another Rs 4,000 crores of additional local environmental damages (that arise largely from the health costs of diesel self-generation). As shown in the figure below, the only significant beneficiaries of stalled reform are the electricity pilferers (i.e. those who steal electricity, which in many Indian States may account for up to 35-45 percent of total generation)

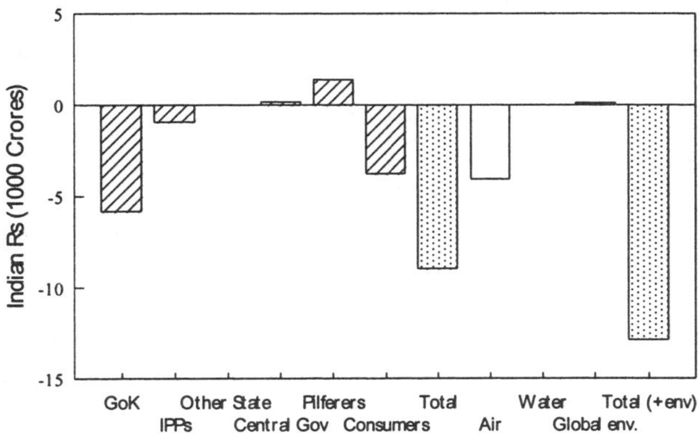

Note: 1 crore (India) = 10 million

Mitigation Options

We next analyse the impact of individual mitigation options. Each option is assessed one at a time as a perturbation of the reference (reform) case. The goal is to derive a cost of avoided carbon for individual options, and to compare the cost effectiveness of each. Based upon this screening process, GHG mitigation scenarios are constructed. In these scenarios, individual options are combined to generate a GHG mitigation option supply curve that takes into account the interactions among individual programmes. The focus is on establishing the economic costs,[12] and the magnitude of the physical GHG emission reduction benefits – which we express as the economic cost of avoided carbon.

Many of the alternative energy options have upward sloping supply curves, which is attributable to two main reasons. The first applies to situations where capital costs are likely to be relatively constant across individual projects, but where energy output (as quantified by annual plant factors) between the best and worst sites shows large differences (for example, as in the case of wind power). The second applies to the converse of this – where plant factors might show relatively small differences, but where construction costs may show high variation (for example, as in the case of mini-hydro, where civil works costs across individual projects show larger variations than run-off coefficients which determine annual load factors).

Alternative Energy Options

Small hydro is an excellent starting point for the assessment of mitigation options, for it is already a component of the IDA/Global Environment Facility Energy Services Delivery Project (ESDP) (World Bank, 1997a). Indeed, grid-connected mini-hydro is its largest single component (US$ 14.4 million of US$ 24.2 million). Four such small hydro plants were connected to the CEB grid in 1996 and 1997: Hydrotech Lanka Dickoya (Pvt) (960 kW); Talawakelle Tea Factory (112 kW); Ritigala Tea Factory (112 kW) and Seethaeliya Tea factory (108 kW).

The mini-hydro potential of Sri Lanka was assessed by Fernando (1998). Small hydro is defined by this assessment as plants of less than 5 MW. A total of 257 sites were identified, broken down as indicated in Table 9.8. Fernando assumes an average capital cost of US$ 1,500/kW of installed capacity at 1998 price levels.[13] He concluded that some 45 MW of the estimated potential could run at annual plant factors of 50 per cent, thereby generating around 200 GWh/year (Fernando, 1998, p. 30).

Table 9.8 Classification of small hydro sites

Site	Status	Sites	Utilized capacity (kW)	Exploitable potential (kW)
Old estates	Abandoned	49	3,343	10,367
	Not in operation	14	544	2,555
	In operation	74	2,228	10,746
New estates		71		20,733
Non-estates		47		53,016
Total		*257*		*97,407*

Source: Fernando, 1998, p. 21–24

A number of more detailed feasibility studies were undertaken in 1996 as part of the background work for ESDP.[14] Table 9.9 shows the main results of these studies.[15]

Table 9.9 Results of five mini-hydro feasibility studies (at Rs 55/US$)

	kW	GWh/ year	Annual PLF	IRR	$/kW	Foreign content
Delta Estate: Pupuressa	292	1.45	0.56	20.53	1,069	0.51
Carolina Estate	2,500	14.04	0.64	20.48	934	0.62
Delmar Estate	339	1.75	0.59	24.05	915	0.47
Peakfield Estate: Maskeliya	2,700	12.30	0.52	19.85	908	0.48
Ellapita Ella: Maliboda Estate	580	2.26	0.44	16.30	950	0.65

Based on these data, we divide the mini-hydro potential into two tranches: a first tranche of 10 MW of the larger projects (such as Carolina and Maskeliya), with capital costs of less than US$ 1,100/kW, implemented in 2000–02 (identified in the figures that follow as "mini-hydro I"), and a second tranche of 35 MW of smaller projects in the range of US$ 1,100–US$ 1,500/kW, implemented between 2001 and 2007 ("mini-hydro II").

One of the main constraints to small hydro development is the adequacy of the tariff, and the extent of a capacity credit. In Sri Lanka, the small hydro tariff paid to private sector developers has been the subject of continuing negotiation (and dispute) with the CEB. At present, the announced tariff is based solely on avoided energy costs, and does not take into account any capacity benefit. However, while the capacity benefit may be small, it is unlikely to be zero.

The main argument against a capacity benefit is that, in the short run, the quantity of mini-hydro is so small that CEB's capacity expansion plan is unlikely to be any different, a result that seems to be confirmed by runs of the WASP model, which identifies little or no impact on the optimal expansion plan. We note that capacity and energy credits need to be estimated in the same model – one cannot use WASP for capacity credits, and some other model (such as METRO) for the energy credits.

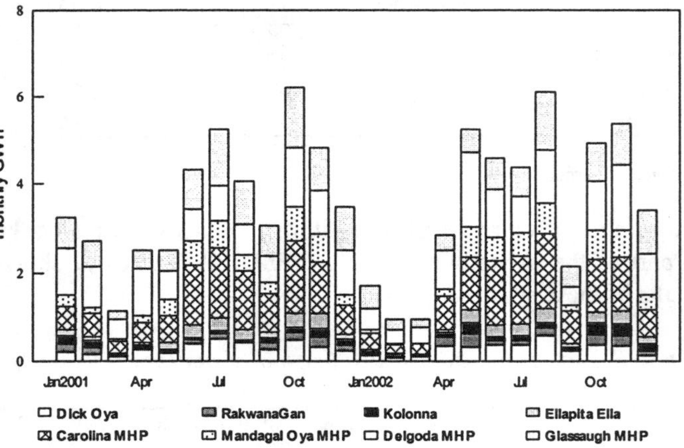

Figure 9.5 Monthly generation from small hydro projects

On the other hand, even in the driest months of February and March, there is still some mini-hydro generation, as shown in Figure 9.5 which summarizes the monthly generation of all the small hydro projects for which we have two years of complete data. As the number of mini-hydro plants increase, it is likely that the diversity effect will increase the capacity value to CEB. In other words, while the capacity benefit of any single plant may be small, because of the monthly and annual variation of a particular site, when the outputs of all small hydro projects are added together the extent of variation declines.

In some locations, while individual technologies cannot be diversified, combinations of different renewable technologies may provide diversification, and increase the capacity credit for the renewables portfolio as a whole. For example, as shown in the Annex, just such a combination may be possible in the Eastern China province of Zhejiang, where small hydro peaks in summer, and wind in winter.

Wind Power: Sri Lanka's wind climate is characterized by two monsoon systems – the southwest monsoon lasting from May to early October, and the northeast monsoon, from December to late February. Southwesterly winds are stronger and also display a greater directional persistence than the northeasterly winds. With strong to moderate winds prevailing for about eight months of the year in some parts of the country, Sri Lanka can be considered a country with a modest potential for wind power generation. A 3 MW wind power demonstration is underway with the support of the ESD project, with an estimated capital cost of US$ 1,175/kW.

Sri Lanka lacks historical wind data of good quality to make a reliable assessment of the wind energy potential. From the viewpoint of wind energy assessment, existing wind data at meteorological stations suffer from several shortcomings, including inadequate height of measurement (typically only 3–6 m), inappropriate frequency of measurement (typically three-hourly during day and none in night time), poor exposure of the anemometer and lack of historical information on instrument calibration, replacement, and change of measuring locations.

Nevertheless, some tentative conclusions regarding the classification of windy regions in Sri Lanka could be drawn from existing wind data and also natural indicators. The situation is much better in the southern and southeastern lowland region where the Ceylon Electricity Board (CEB, 1992) conducted a three-year wind-monitoring programme covering about 1,500 km^2. Data gathered in this programme consist of hourly average wind speed and maximum wind speed for each hour at 10, 15 and 20 m heights, and wind direction at 20 m. Table 9.10 shows the exploitable wind power: this analysis does not take into account the Central and Knuckles mountain regions due to the uncertainties associated with wind power prediction in complex terrain without firm data and modelling work.

Table 9.10 Wind power potential

	Approximate size of the region			Fraction	Area	Potential	
	Length (km)	width (km)	km^2	(%)	(km^2)	(MW/ km^2)	MW
Mirijjawila–Hambantota	12	1	12	50	6	6	36
Karagam lewaya	3	1.5	5	50	2	6	14
Kirinda–Palatupana	12	1.5	18	50	9	6	54
Weligatta–Weerawila	6	6	36	50	18	6	108
Sevanagala sugar plantation	6	6	36	50	18	6	108
Suriyawewa–Mrijjawila	12	1.5	18	50	9	6	54
Puttalam–Dutch Bay	40	1	40	50	20	6	120
Mannaer Region	40	1.5	60	50	30	6	180
Ponnaweli–Pooneryn		1.5	75	50	38	6	225
Jaffna lagoon	25	6	150	50	75	6	450
Manalkadu–Chundikulam	50	3	150	50	75	6	450
Total potential			600		300		1,799

Detailed studies using the WASP model were undertaken to examine the maximum amount of wind power that could be absorbed by the system while still meeting the planning criteria established by the CEB (LOLP of less than 0.1 per cent in the years after 2001). These studies reveal that as much as 2,000 MW – that is, slightly more than the 1,800 MW indicated in Table 9.10 (that reflects 50 per cent development of the respective areas) – could be absorbed into the system over the planning horizon.

This maximum wind power scenario is shown in Figure 9.6. We assume a decrease in capital costs from the present US$ 1,175/kW to US$ 1,000/kW over the planning horizon. Significantly lower capital costs may be achievable if the rate of progress in wind turbine technology achieved over the past decade continues into the future.[16]

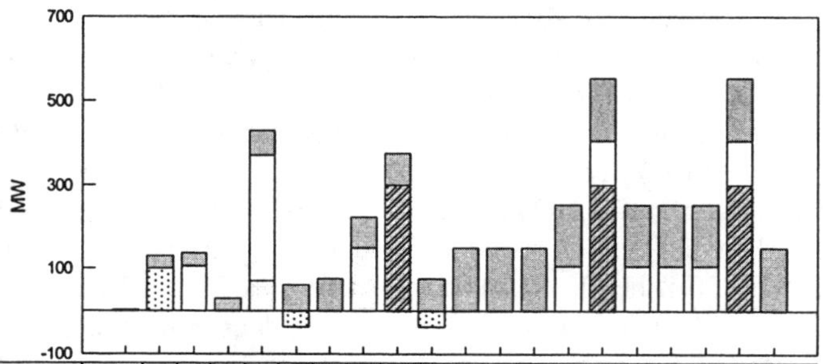

	1999		2001		2003		2005		2007		2009		2011		2013		2015		2017	
hydro					70			150												
coal									300						300				300	
liquid fuels			106		300									105	105	105	105	105	105	
gas																				
diesel		100				-36				-36										
renewables	3	30	30	30	60	60	75	75	75	75	150	150	150	150	150	150	150	150	150	150

Figure 9.6 Capacity expansion plan, maximum wind development

Such scenarios for renewable energy, and their assumptions, need much caution. In addition to the year-on-year variation in wind speeds (discussed in Chapter 4), there is the additional uncertainty in the estimation of the long-term average wind speed, and hence the long-term average annual production. For example, in the case of wind projects in Denmark, for which an excellent database exists for over 1000 different sites, actually achieved wind energy production over the last decade is considerable smaller than that expected at design (Box 9.2).

Dendro-thermal power: There are many proposals for dendro-thermal power plants for Sri Lanka (Wijeratne and Joseph, 1999). The main rationale is their renewable nature, avoiding costly fossil fuel imports. Equally important is employment generation in areas presently characterized by high unemployment.

Box 9.2 Forecasting wind energy production: the Danish experience

The Danish Windstats database contains extensive data for all wind turbines presently operating in Denmark. Based on 1,080 sites for which there are more than six years of wind annual production data, the actual average annual production is 12 per cent lower than predicted at the time of wind farm design. Figure A plots the shortfall (actual average MWh divided by design MWh) as a function of the start year: the more recent the wind farm, the greater the extent of over-estimation.

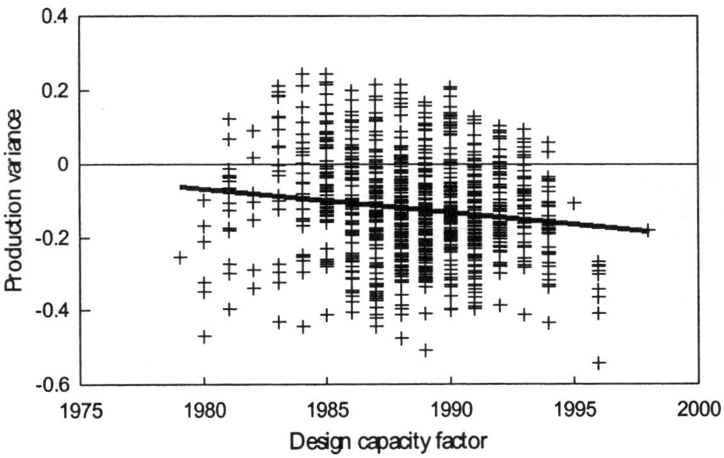

Figure A

There are three hypotheses that could explain this trend: first, that developers and turbine manufacturers have indeed become more optimistic over time. Secondly, that machines have become significantly more unreliable, though there is little evidence of this being so in the case of Danish wind farms. The Windstats database reveals very few hours lost due to maintenance problems: in 2000, the estimated loss of output attributed to this cause was less than 1 per cent). And thirdly, (and arguably the most likely) that there has been a fundamental change in wind regime over the past decade. Indeed, for the national average, wind speeds for all six years 1995–2000 appear to have been significantly below the predicted long-term average. Yet if changes of wind regime occur over such relatively short time horizons, then wind power is subject to a significant additional risk factor.

We assess the potential of dendro-thermal power plants for GHG mitigation by postulating a scenario in which baseload coal plants are replaced by dendro-thermal units in 50 MW increments.[17] This results in the capacity expansion plan indicated in Figure 9.7.

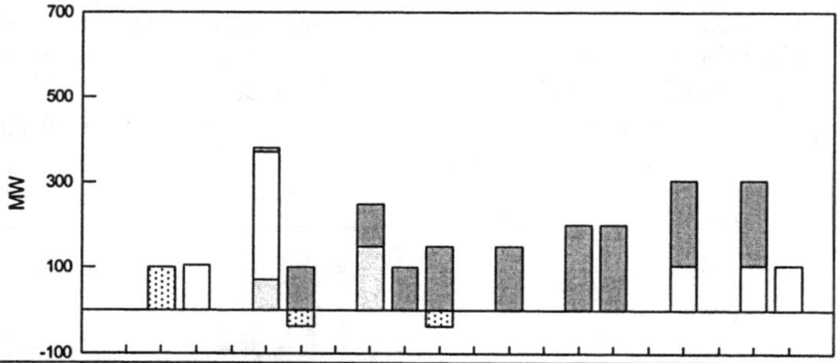

	1999		2001		2003		2005		2007		2009		2011		2013		2015		2017	
hydro					70			150												
coal																				
liquid fuels			106		300												105		105	105
gas																				
diesel		100				-36				-36										
renewables					10	100	0	100	100	150		150		200	200		200	0	200	0

Figure 9.7 Capacity expansion plan for the dendro-thermal scenario

It is assumed that gasifier rather than combustion (steam cycle) technology is employed: the assumptions are given in Table 9.11.

Table 9.11 Assumptions for dendro-thermal power plants

		Dendrothermal (gasifier)	Coal plant
Unit size	(MW)	10	2 × 300
Capital cost	(US$/kW)	1,200	899
Fixed O&M	(US$/kW/month)	1.8	0.64
Var O&M (non-fuel)	(US$/MWh)	2.5	3.54
Heat rate	(Kcal/kWh)	4,560	2,162
Sulphur content	(%)	0.02	0.6
Heat value as received(a)	(Kcal/kg)	3,800	6,300
Plant gate cost	(US$/ton)	22	48
Land area for fuelwood plantation	(Ha/MW)	360	

Note: (a) assuming 20 per cent moisture for fuelwood

Source: Based on information provided by P. G. Joseph of the Sri Lanka Energy Conservation Fund

Dendro-thermal power plants require substantial land for the necessary fuel-wood plantations. However, the advocates note that 1,000 MW of dendro-thermal would require only some 20 per cent of present scrub and chena lands (Table 9.12).

Table 9.12 Land requirements for dendro-thermal plants

Total land area of Sri Lanka (ha)	6,500,000
Scrub lands (ha)	600,000
Chena lands (ha)	1,000,000
Land requirement for: 10 MW (ha)	3,600
1,000 MW (ha)	360,000

A more potent question concerns the fuel cost (assumed at US$ 22/tonne, delivered to the plant gate). This is based upon the present price of fuelwood as delivered in Colombo. Fuelwood produced in properly managed plantations, with power plants sited at the centre of such plantations, may reduce this cost: to test the sensitivity of the resulting cost of avoided carbon to fuel cost assumptions, we therefore consider two dendro-thermal fuel-cost scenarios, one at US$ 22/tonne, the second at US$ 14/tonne.[18]

Solar PV for remote rural applications: Solar photovoltaic systems for remote rural homes ('solar homes') represent another component of the ESD project (whose economics are discussed further in Chapter 11). Such systems have been marketed for some time by commercial vendors, but Global Environment Faculty (GEF) support is warranted by the many institutional and financing constraints that have impeded greater market penetration to date.[19]

In the solar homes scenario we assume that by 2018, about 85 per cent of the then remaining unelectrified homes would be served by solar systems, representing about 500,000 systems in place. This is significantly greater than the present estimate of between 100,000 to 215,000 homes considered by private vendors as the present potential market, taking into account affordability criteria (see, for example, Jayewardene and Perera, 1991). However, over the 20-year planning horizon considered here, and given the GDP growth projection also used as a basis for electricity demand forecasting, a doubling of real income (and, say, a consequent doubling of the potential market) ought to be achievable. Moreover, we assume that the present cost of US$ 6,000/kW will decrease over time, reaching US$ 1,000/kW by 2018, and the unit output per system would increase from the present 30–50 watts[20] to 200 watts by 2018. These postulated improvements (as well as the market penetration assumptions) may well be optimistic, but seem warranted in order to determine the bounds of potential cost-effectiveness for GHG mitigation.

Fuel Substitution Options

Oil steam-cycle plants: For this GHG assessment we assume that 3.5 per cent high sulphur fuel oil would be imported from the Gulf, and that flue gas desulphurization (FGD) would be required (as uncontrolled emissions would exceed standards, and are higher than from plants burning Australian low-sulphur coal). The plant characteristics are taken from Electrowatt (1996).

LNG: The economics of LNG were discussed in Chapter 5. Here we take the LNG price, delivered to the plant gate, at US$ 5.00/mmBtu. This includes the cost recovery of the up-front infrastructure costs (terminal, gas storage, regasification facility, etc.) Capital and operating costs of the power generation plant, heat rate, etc., may be taken as the same as for diesel-fuelled combined cycle plants.

Conventional hydro: Among larger conventional hydro schemes, only the Kukule and Upper Kotmale (Talawakelle) projects are in the least-cost generation expansion plan, and consequently to examine the effect of additional hydro as a GHG mitigation option, the remaining hydro projects have to be forced into the solution, with exogenously specified commissioning dates.[21] This maximum hydro development scenario is shown in Table 9.13 (and is identified in the figures as "MaxHy").[22]

Table 9.13 Additional hydro projects for the maximum hydro scenario

	Start-up year	MW	Capital cost ($/MW)(a)
Broadlands	2005	40	2,548
Gin Ganga	2005	49	2,127
Moragolla	2007	27	3,002
Uma Oya	2010	150	2,152

Note: (a) pure (overnight) project cost, without IDC, customs, duties

Biomass co-firing: Given the high cost and uncertainty surrounding dendro-thermal gasifiers, biomass co-firing at the proposed coal-fired plants may be possible. Supplying 15 per cent of the energy input from fuelwood should be feasible at minimal modification of the coal plant (about US$ 75/kW). Table 9.14 summarizes the assumptions used for this case, which are based on the EPRI Technical Assessment Guide. The plantation fuelwood as might be used in Sri Lanka would have a higher heat value and lower moisture content than the wood fuel blend (of urban wood, mill, forest and agricultural crop residues) assumed in the EPRI study.

Table 9.14 Assumptions for wood co-firing

	EPRI TAG Assumptions	As used in this report
Power consumption in wood pulverizers	60 kWh/dry ton	
Reduction in boiler efficiency	1.5%	
Wood characteristics	33% moisture, 5,495 Btu/lb=3,022 KCal/kg	20% moisture 3,800 KCal/kg
Hence: increase in net plant heat rate	166 Btu/kWh	45 KCal/kWh
Wood pulverizers	US$ 19/kW (at 1993 prices)	US$ 25/kW
Fuel handling facilities	US$ 31/kW (at 1993 prices)	US$ 40/kW
Modifications of boilers and other facilities	US$ 6/kW (at 1993 prices)	US$ 10/kW

DSM: A programme of lighting DSM is assumed included in the reform case, as described above. For the further GHG mitigation cases we separately examine energy efficient motors (EEM) and energy efficient air conditioning (EEAC). Table 9.15 shows the results of the screening analysis of individual options, listed in order of avoided cost.

Table 9.15 Screening of DSM programmes

	US$/ton CO_2 Avoided	Avoided carbon, 1000t	
		Measure	Cumulative
IND:slimTube	-51.4	353	353
COMM:slimTube	-48.3	693	1,046
RES:23W CFL	-46.0	481	1,527
IND:EEM	-41.9	403	1,930
RES:18W CFL	-31.3	525	2,455
IND:incan>18W CFL	-23.8	412	2,867
RES:13W CFL	-14.2	1,181	4,048
COMM:EEAC	-10.2	417	4,465
COMM:incand>Slcomfort	2.2	779	5,244

Note: negative values are 'win–win'

9.2 RESULTS

The results of the screening analysis are summarized in Table 9.16. Note that the reductions in emissions in columns [4] and [5] cannot be added, since the individual data points represent perturbations of the reform scenario, one at a time (and some are mutually exclusive, such as LNG, oil steam-cycle with FGD, or the dendro-thermal options at different delivered wood costs). Negative costs of avoided carbon indicate 'win–win options'.

Table 9.16 Cost of avoided carbon

	System cost $million	δ (cost) (relative to reform)	CO_2 emissions (million tons)	δ (CO_2) (relative to reform)	δ (C)	Cost of avoided carbon $/ton C
	[1]	[2]	[3]	[4]	[5]	[6]
Reform	2,449		98.4			
+DSM:ind:EEM	2,430	-18.5	96.6	-1.8	-0.5	-37.6
Oil Steam (Resid)	2,325	-123.7	83.1	-15.3	-4.2	-29.7
Oil Steam (Resid)+FGD	2,362	-87.2	83.1	-15.3	-4.2	-20.9
+DSM:CommAC	2,441	-7.6	96.4	-2.0	-0.6	-13.6
Oil Steam+FGD	2,416	-33.2	88.5	-9.9	-2.7	-12.3
+T&D>10%	2,440	-9.2	95.2	-3.2	-0.9	-10.4
Mini-hydro:I	2,448	-0.8	97.5	-1.0	-0.3	-3.2
Mini-hydro:II	2,450	1.5	95.7	-2.7	-0.7	2.1
Dendro-thermal, 14$/t	2,492	42.7	39.2	-59.3	-16.2	2.6
Oil Steam (0.3%S)	2,465	16.5	88.6	-9.8	-2.7	6.1
Dendro-thermal, 22$/t	2,580	131.1	39.6	-58.8	-16	8.2
Diesel ($800/kW)	2,494	45.3	83.8	-14.6	-4	11.4
LNG	2,545	96.3	77.3	-21.1	-5.8	16.7
Diesel ($1,140/kW)	2,580	130.9	83.8	-14.6	-4	32.8
MaxHy	2,553	104.0	88.3	-10.1	-2.8	37.7
Wind (Max)	2,870	421.2	74.8	-23.7	-6.5	65.3
Solar Homes	2,490	41.6	97.6	-0.8	-0.2	184.5

Figure 9.8 displays these data in a different way, plotting system cost against GHG emissions (undiscounted, as CO_2), with the four quadrants defined by the reform case. The slope of the line connecting any point with the reform case defines the cost of avoided carbon (column 5 of Table 9.16).

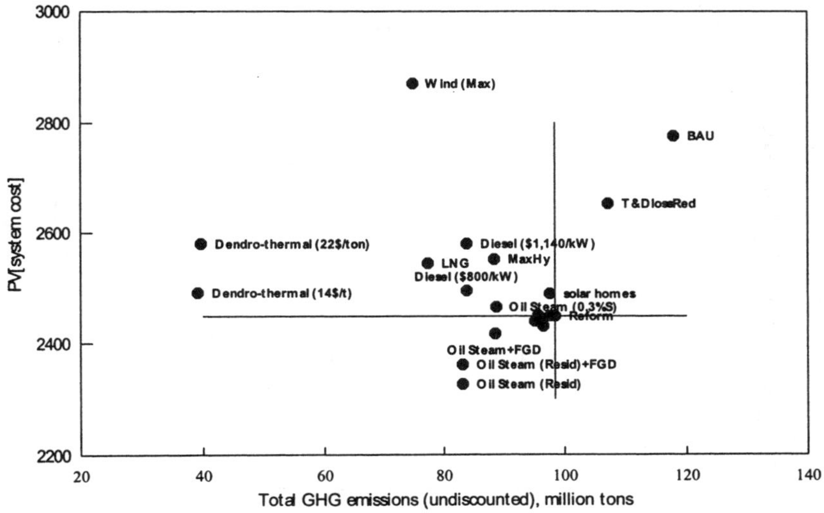

Figure 9.8 System cost versus GHG emissions

There are few surprises in these results. The most expensive in terms of the cost of avoided carbon is the solar homes programme at US$ 184/tonne of carbon. However, this result is a good example of the hazards of conducting the analysis in terms of changes in system costs rather than changes in net benefits. As we shall see in the detailed analysis of Chapter 11, when benefits of providing lighting and TV-viewing in off-grid areas are properly assessed, the solar home option brings significant net benefits.

Impact on Local Air Emissions

Implementation of GHG emissions reduction measures also reduces local air emissions (SO_2, NOx, particulates). Figure 9.9 shows the relationship between GHG emissions and SO_2 emissions for the individual mitigation measures (expressed as perturbations of the reference scenario). However, the monetized value of these local environmental benefits is small in comparison to the cost of avoided carbon.

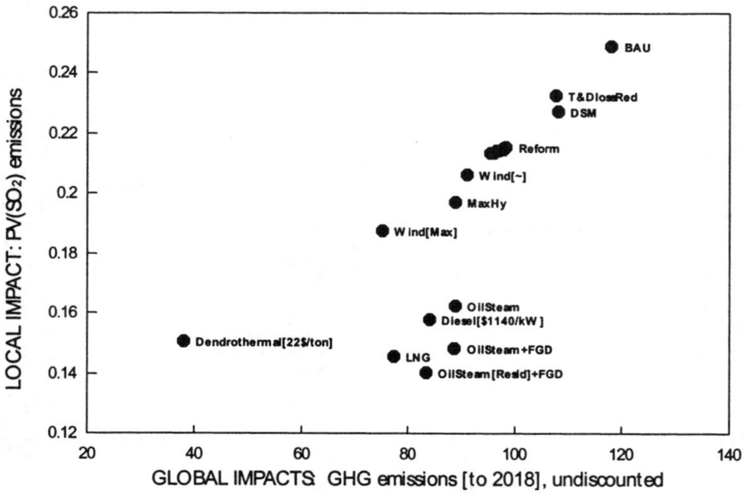

Figure 9.9 Correlation between GHG emissions and SO_2 emissions

GHG Emission Reduction Scenarios

Starting with BAU, individual mitigation measures are brought into the solution cumulatively. The criterion for introducing the next measure could be the next lowest cost of avoided carbon, or the measure most likely to be implemented. For example, starting with BAU, the first step could be either DSM (which has the best cost of avoided carbon) or T&D rehabilitation (which experience in India shows is more easily implemented as the first step of reform). We show in Figure 9.10 T&D loss reduction as the first step.

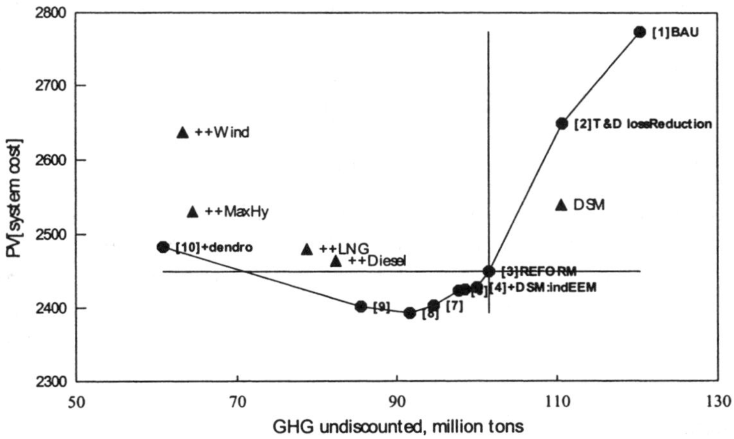

Figure 9.10 Defining the GHG mitigation scenario

Table 9.17 shows the order in which the measures are brought into solution, reaching the least-cost scenario (step [8]) after introduction of all of the DSM, mini-hydro and further T&D loss reduction.[23] Any further measures to reduce GHG emissions would then begin to increase costs.

Table 9.17 Definition of mitigation scenarios

	Measures								mutually exclusive options						
	CEB:base case	T&D >12.5%	DSM lighting	DSM Ind EEM	DSM Comm AC	MiniHyI	T&D >10%	MiniHyII	Oil Steam +FGD	Dendro	Diesel	LNG	MaxHy	Wind	Solar Homes
Cost of avoided carbon, $/ton C >		-82	-35	-50	-20	-16	-14	-0.5	-21	8	11	16	38	66	150
Scenarios															
1 BAU	X	X													
2 +T&D>12.5%	X	X													
3 Reform	X	X	X												
4 +DSM:IndEEM	X	X	X	X											
5 +DSM:CommAC	X	X	X	X	X										
6 +miniHydro:I	X	X	X	X	X	X									
7 +T&D>10%	X		X	X	X	X	X								
8 +miniHydro:II	X		X	X	X	X	X	X							
9 Oil-Steam[+FGD]	X		X	X	X	X	X	X	X						
10 Dendrothermal	X		X	X	X	X	X	X		X					
11 Diesel	X		X	X	X	X	X	X			X				
12 LNG	X		X	X	X	X	X	X				X			
13 +MaxHy	X		X	X	X	X	X	X		X			X		
14 +Wind	X		X	X	X	X	X	X		X			X	X	
15 +SolarHomes	X		X	X	X	X	X	X		X			X	X	X

By the time we reach the economic least-cost scenario (all the DSM, mini-hydro and further reduction of T&D losses) – the lowest point on the curve – the number of coal plants over the planning horizon has reduced further from the five in the reform scenario to four.

Developing the Standard Abatement Cost Curve (SACC)

Table 9.18 shows the avoided costs of carbon and the percentage reductions (relative to the BAU case) achievable in each scenario.

It may be noted that win–win options, when considered as a stand-alone perturbation of the reform case (such as OilSteam+FGD), may not necessarily still be win–win when considered together with other win–win options in a scenario. As shown in Table 9.18, introducing Oil Steam+FGD *after* the last tranche of mini-hydro increases costs, and is therefore no longer win–win.

Table 9.18 GHG mitigation scenarios

Scenario	System cost	δ (cost)	CO_2	δ (CO_2)	$/ton C	% reduction (BAU)
BAU[#]	2,775	382	118.0			
T&DlossRed	2,653	260	107.1	19	-49.5	9
Reform	2,449	56	98.4	11	-19.4	17
DSM:ind:EEM	2,430	38	96.6	9	-15.7	18
DSM:CommAC	2,425	32	95.3	7	-15.7	19
Mini-hydro I	2,423	30	94.6	7	-16.2	20
T&D>10%	2,403	10	90.5	3	-13.8	23
Mini-hydro II =least-cost	2,393		87.8			26
OilSt+FGD	2,402	10	79.6	-8	4.4	33
Dendro	2,481	89	46.2	-42	7.8	61
Diesel	2,500	107	75.4	-12	31.6	36
LNG	2,465	73	70.2	-18	15.2	41
MaxHy	2,540	147	49.8	-38	14.2	58
Wind	2,639	246	50.1	-38	23.9	58

The corresponding GHG mitigation supply curve – where we introduce each measure in order of avoided cost – is shown in Figure 9.11. Here we draw the SACC in tonnes of carbon (rather than tonnes of CO_2)

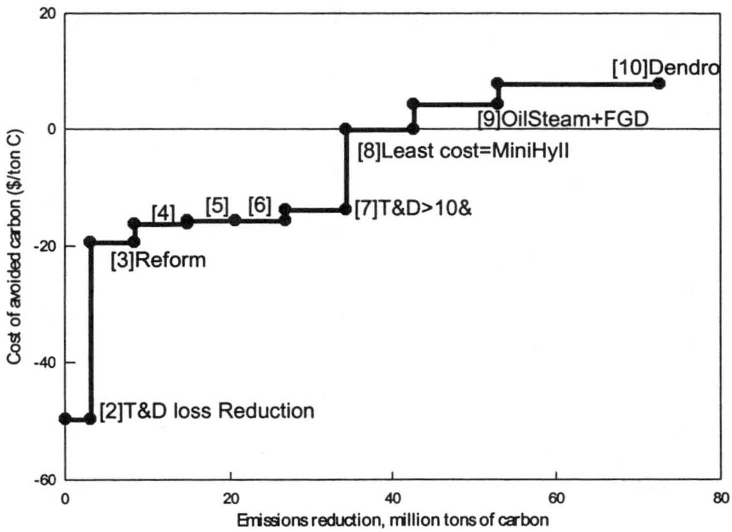

Figure 9.11 Standard abatement cost curve

In a standard deterministic framework, all the steps that lie below the emission-trading price should be pursued if decision-makers seek to maximise the net benefits of abatement. But when the uncertainties are properly considered in a dynamic framework, things are not that simple (as we shall see in Chapter 10).

The Sri Lankan results may be compared to those of the two GHG overlays prepared for the Indian States of Andhra Pradesh (ASCI, 1999) and Bihar (SCADA, 1999) (Table 9.19). Our results are similar to the results in India for some technologies (lighting DSM, further reduction of T&D losses), but quite different for others (such as LNG). These results are all expressed in terms of undiscounted emissions.

Table 9.19 Comparison with Indian GHG overlay study results: cost of avoided carbon in US$/tonne of carbon (measured against reform scenario)

	Sri Lanka	Andhra Pradesh	Bihar
DSM	-13 to –38 (EEM, AC)	-28.9 (lighting)(a)	-18
DSM-agriculture		-14.3	
Biomass	8.2 (dendro-thermal)	-18 (bagasse)	-44 (bagasse)
Refinery residue		3.3	
T&D rehabilitation	-10.4	-19.4	-36
LNG	17	0.7	24.1(b)
Thermal plant rehabilitation(c)		-71.5	
Mini-hydel	-3.2 to 2.1	4	-21.1
Large hydro	38	-8.4(d)	34
Nuclear		9.5	2.5
Wind	65	22	
Solar photovoltaic	184	100	108
Coal washing		167	

Notes:
(a) In Sri Lanka, lighting DSM is assumed to be part of the reform programme
(b) Pipeline gas imported from Bangladesh at US$4/mmBTU
(c) In Andhra Pradesh: rehabilitation of the Kothagudem thermal plant; in Bihar, part of baseline reform scenario; not applicable to Sri Lanka
(d) The hydro in AP is the powerhouse-only plant at Jurala, at an existing irrigation project
(e) Indian results are assessed and reported as Rs/tonne, converted to US$ at Rs 43/US$

The LNG results in the case of Andhra Pradesh – which suggest a cost of avoided carbon of only US$ 0.7/tonne rather than US$ 17/tonne as in Sri Lanka[24] – are particularly interesting, since, as in Sri Lanka, LNG is assumed to replace imported Australian coal.[25] However, as is evident from Table 9.20, the Andhra Pradesh results are based upon a lower LNG price (the difference of US$ 0.50/mmBTU reflects the difference in scale, Indian LNG developments allowing 2,000 MW plant sizes), and much higher costs for imported coal plants:[26] consequently the cost of avoided carbon in Andhra Pradesh is much lower.

Table 9.20 Comparison of assumptions, Andhra Pradesh v. Sri Lanka

	Andhra Pradesh	Sri Lanka
Coal		
Capital cost	US$ 1,139/kW	US$ 899/kW
Fuel cost	US$ 2.16/mmBTU	US$ 1.85/mmBTU
LNG		
Capital cost	US$ 697/kW	US$ 634/kW
Fuel cost	US$ 4.50/mmBTU	US$ 5.00/mmBTU

9.3 THE IMPLICATIONS OF ALTERNATIVE TRANSPORT SCENARIOS

The relationship between vehicle ownership and per capita income differs from country to country.[27] In other words, the rates of vehicle ownership of two countries when they had the same income levels are observed to be quite different. The work of Newman and Kenworthy (1990) presents some insights on the parameters that appear to significantly influence the level of vehicle ownership. These can be identified as:

- Population density
- Penetration by public transport
- Cost of vehicles relative to income level
- Taxation levels on private vehicles and
- Restraint on vehicle ownership.

It is observed that in countries that have lower population density, penetration by public transport is difficult, and so these countries will tend to have higher vehicle ownership rates compared to others. In this respect, countries in Asia appear to have lower vehicle ownership rates when compared to countries such as the UK, Australia and Canada. Figure 9.12 shows a comparison of international data for car and van ownership as a function of per capita income.

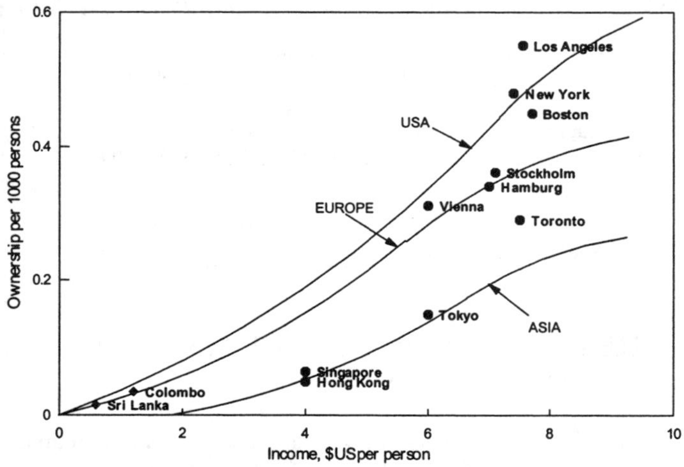

Source: Kumarage (1999a), Figure 2

Figure 9.12 Vehicle ownership trends and likely saturation levels as a function of income

India, Sri Lanka and other developing nations have historically levied high taxation on the ownership and use of private vehicles and thus the growth rates have been further reduced. Strong restraints of vehicle ownership by the government have recently been introduced in countries such as Singapore and Hong Kong – but these differ fundamentally to the restraints imposed by the erstwhile socialist governments of Sri Lanka. The new generation of constraints is based on market-based instruments to allocate shortages – in Singapore's case by the auctioning of a limited number of permits and through road pricing (whose revenues are used to improve public transportation) – while in the old days of socialist import controls in Sri Lanka, bureaucratic discretion was the instrument of allocation.

Ownership trends need to be studied for different vehicle types. Two and three-wheelers (that is, motorcycles and 'tuk-tuks' or 'trishaws' as they are called in different countries), which are popular modes of private transport in Asia, have patterns quite different to four-wheeled private vehicles made up of cars, saloons, station wagons, double cabs and passenger vans. Commercial vehicles made up of trucks, delivery vans, buses, taxis and agricultural vehicles have yet different trends.

The environmental implications of increasing vehicle ownership rates are no less profound. In Figure 9.13 we show GHG emissions from the Sri Lanka transport sector for three possible futures (under a high economic growth scenario). The difference in 2020 GHG emissions between the 'American' future and the 'Singapore' future is a factor of three.

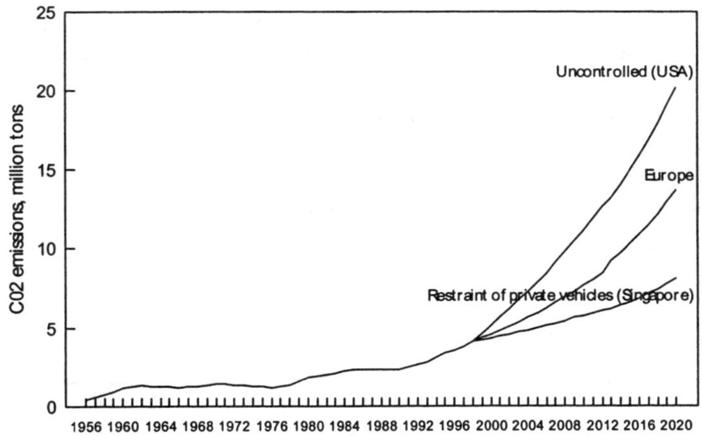

Figure 9.13 GHG emissions: high economic growth

Table 9.21 shows 2020 GHG emissions, including power sector emissions from above, under the three transportation futures as a function of economic growth rate.

Table 9.21 Year 2020 CO_2 emissions

		USA	Europe	Singapore
GHG emissions				
High (6.5%)	(million tons)	47.2	40.6	36.5
Medium (5%)	(million tons)	35.8	31.7	29.1
Low (4.5%)	(million tons)	32.1	28.6	26.5
GDP@PPP				
High (6.5%)	($USbillion)	223	223	223
Medium (5%)	($USbillion)	163	163	163
Low (4.5%)	($USbillion)	147	147	147
kgCO_2/$GDP(@PPP)				
High (6.5%)	(kg/$GDP)	0.21	0.18	0.16
Medium (5%)	(kg/$GDP)	0.22	0.19	0.18
Low (4.5%)	(kg/$GDP)	0.22	0.19	0.18
GHG emissions per capita				
High (6.5%)	(kg/capita)	1.81	1.56	1.40
Medium (5%)	(kg/capita)	1.38	1.22	1.12
Low (4.5%)	(kg/capita)	1.23	1.10	1.02

In this table, the numbers in brackets represent annual GDP growth rates: in the year 2020, these compute to US$ 8,572/ capita for the high GDP growth case (6.5%), US$ 6,274 for the medium growth case (5%), and US$ 5,669 for the low growth case (4.5%).

While these GHG emissions show significant increases over their present levels, when expressed in terms of emissions per capita, or emissions per unit of GDP, Sri Lanka will still be far below the present world averages assuming a scenario midway between the Europe and Singapore futures (Table 9.22). Even under the most pessimistic standpoint from the point of view of increasing road congestion (that is, high economic growth), at least GHG emissions would not increase to a point where Sri Lanka's ability to benefit from carbon offset trading would be compromised.

Table 9.22 International comparisons

	GHG/ capita	$kgCO_2$/ $GDP	$kgCO_2$/$GDP PPP adjusted
1995			
Sri Lanka	0.34	0.63	0.13
China	2.51	4.72	0.92
Germany	10.83	0.49	0.63
France	6.23	0.29	0.34
India	0.86	2.17	0.73
Indonesia	1.17	1.37	0.34
Japan	9.17	0.36	0.46
Pakistan	0.63	1.62	0.32
USA	19.88	0.85	0.85
Former USSR	8.48	5.22	2.61
World average	3.92	0.94	0.75
2020 Sri Lanka			
High economic growth (6.5% per annum)	1.35–1.8		0.16–0.21
Medium (5%)	1.09–1.38		0.17–0.22
Low (4.5%)	1.00–1.23		0.18–0.22

ANNEX:
THE DIVERSITY BENEFITS OF RENEWABLE ENERGY PORTFOLIOS

Diversity benefits in renewable energy portfolios have been examined in China (Meier, 2004). Figure 9.14 shows the relative monthly output of the representative small-hydro projects in Zhejiang province compared to the seasonal variation in demand for the Zhejiang grid as a whole.

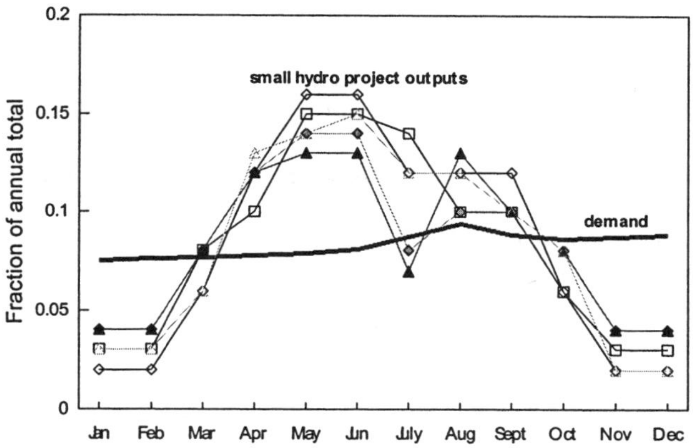

Figure 9.14 Seasonality of small hydro production in Zhejiang Province

It is evident that the seasonality of output and grid demand is poorly matched, even in the case of small hydro projects with substantial storage. Figure 9.15 shows the corresponding data for the seasonality of wind-farm generation. Again we see a poor match between seasonal output and grid demand.

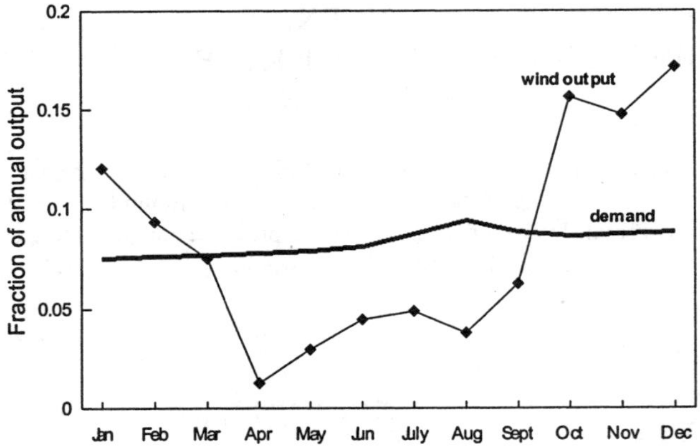

Figure 9.15: Demand and monthly output of Zhejiang wind farms

But when the output of the two types of renewable energy project is combined, one obtains the result of Figure 9.16, now reasonably well-matched to the annual demand curve.

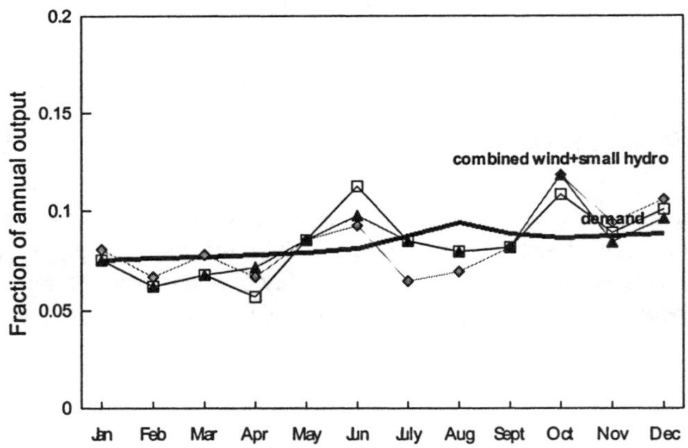

Figure 9.16 Sum of output of small hydro and wind farms

Moreover, the result shown in Figure 9.16 assumes no change in operating rule at those small hydro projects that do have some degree of seasonal storage. While changes in operating rule at multi-purpose projects (25 per cent of the total in Zhejiang) may be difficult, and at pure run-of-river projects not possible at all, some 35 per cent of small hydro projects in Zhejiang have sufficient storage capacity to permit optimisation of the operating rule. In order for this combination to provide real benefit, the small hydro plants must have at least daily peaking capability, since wind farm output rarely matches the daily load curve.

NOTES

1 Meier and Munasinghe (2003c) discuss the question of whether different power system planning models yield different conclusions: they do not. Differences in models prove to be quite minor compared to different assumptions.

2 Including by BCHydro (1995) for strategic planning and to support stakeholder consultations; by USAID for the 1995 Integrated Resource Plan of the Indian State of Andhra Pradesh (USAID, 1995), by the World Bank in 1997 and 1998 for the economic appraisal of power sector reform programmes in the Indian States of Haryana (World Bank, 1997b) and Andhra Pradesh (World Bank, 1997e), and by the World Bank in a study of environmental issues in the Indian power sector (World Bank, 2004a, 2004b).

3 A good example of why such an approach is desirable (and necessary) is flue gas desulphurisation (FGD), the need for which has been debated in Sri Lanka every time a new proposal for a coal-fired power project has been advanced. Plants fitted with FGD systems will have lower overall efficiency (due to higher auxiliary consumption and impact on boiler efficiency), and may therefore *increase* CO_2 emissions per net kWh by up to 5 per cent.

We might note parenthetically that in addition to the efficiency effect, SO_2 removal, whether by lime or limestone FGD systems, or by fluidized bed combustion systems where calcination occurs in situ, involves the production of CO_2 as indicated by the (greatly simplified) summary reactions $CaCO_3 \rightarrow CaO+CO_2$; $CaO+SO_2 +0.5O_2 \rightarrow CaSO_4$. These chemical reactions imply that for every mole of SO_2 reduced, one mole of CO_2 is produced. However, in the case of a coal containing 1 per cent sulphur by weight (which would be at the upper limit of sulphur content of typical export coals), the contribution of CO_2 from these FGD reactions is about 1.5 per cent of the CO_2 produced by combustion.

4 Since the power sector does not emit CFCs and HCFCs, these are omitted. Carbon monoxide, non-methane hydrocarbons and nitrogen oxides (other than nitrous oxide) are shown in the GHG Overlay Guidelines as having zero (or negligible) GWP and hence are also omitted.

5 The econometrically determined forecasting equations for domestic, industrial and commercial sectors are of the autoregressive form $D(t)=a+bD(t-1) +cGDP(t-1) +dGDP(t)$ (where a,b,c,d are coefficients estimated by least squares, with $R^2>0.96$ in all cases).

6 Studies are currently underway at CEB to confirm these estimates.

7 Care will need to be exercised when evaluating pricing options in any bottom-up, partial equilibrium analysis to correct for welfare effects. Elimination of subsidies, reduction of non-technical losses, or imposition of LRMC-based tariffs, implies loss of consumer surplus to some consumer groups and possible gains in producer surplus (Munasinghe and Warford 1980). Clearly if system cost is used as the measure of benefits then this must be adjusted for any changes in benefits associated with any particular option.

8 LECO makes purchases at 11kV, and therefore its loss rate does not include losses in transmission at the higher voltages.

9 However, it may be noted that Shrestha et al. (1998) examined the impact of DSM in a dynamic systems planning model, but used an assumed LRMC as the yardstick for evaluating the cost of conserved energy.

10 Comparisons with Haryana, the second Indian state (after Orissa) to embark on comprehensive power sector reform, are instructive, particularly since Haryana is among the smaller Indian states and comparable in size to Sri Lanka. (World Bank, 1997b). While in Sri Lanka the consequences of reform are economic benefits as well as reductions (about 15 per cent) in GHG emissions, in Haryana, the economic benefits are some six times larger than in Sri Lanka, but reform also results in a 25 per cent increase in GHG emissions. This is a consequence of the massive inefficiencies in the Haryana system (such as T&D losses of 35 per cent, almost twice the Sri Lanka level), and the significant present curtailments (25 per cent of demand), which would be eliminated by reform (by restoring commercial creditworthiness, allowing presently stalled IPPs to come to financial closure).

11 However, when these benefits are monetised, they account for only a small fraction of the carbon offset benefits.

12 All references to 'cost' in this chapter therefore mean economic cost, without taxes and duties, and without interest during construction.

13 This is substantially higher than the US$ 1,030/kW figure used by the ESDP.

14 These studies were prepared by Consultancy and Professional Services (Pvt) Ltd., for the following projects: Peakfield Estate at Maskeliya; Delmar Estate; Carolina Estate; Delta Estate at Pupuressa; and Ellapita Ella. The latter project was used as a basis for the financial analysis in the ESDP project appraisal document (PAD) which is discussed further in Chapter 11. A second study for the ESDP project examined eleven projects (Posch and Partners, 1994), with generally similar results.

15 A number of issues arise from these studies. The price levels and exchange rates are not stated, but we have assumed the average 1996 exchange rates for the calculation of the US$/kW indicated in Table 9.9. Assuming 100 per cent foreign import for the electromechanical and power transmission equipment, we can estimate the foreign content also as shown (though again this is our assumption). There is no distinction between economic and financial analysis, and the IRRs calculated are financial, based on current price levels (so the real IRR would be lower by roughly the assumed inflation rate of 5 per cent).

16 The 1997 EU White Paper on Renewable Energy, for example, projects somewhat optimistically that by 2010, windpower capital costs, including site costs, would fall to €700/kW from the present €1,000–1050/kW.

17 This is purely for modelling convenience. In actuality the increments could be in much smaller units, more closely matching the year-by-year incremental capacity requirements.

18 The analysis of fuelwood plantations using eucalyptus grandis in a seven-year growing cycle by Walpita (1997) shows a financial cost of US$ 21/tonne of fuelwood. However, the short-rotation coppice system shows considerable promise for lowering these costs. Fast-growing coppicing species such as Gliricidia, Leucaena, Casuarina and Felcipium, would be planted at about 1m spacing, and are initially cropped at about a height of 1m. Subsequent harvesting can be twice a year, with an overall biomass yield per hectare about double that of traditional fuelwood trees.

19 The ESD programme will provide US$ 100 grant per solar home module of 30 watts or greater.

20 The ESD programme that targets 2,200 medium income (income greater than Rs 3,000/month) rural households in the Galle district assumes that 30 per cent of the system sold would be 30 watt units, 40 per cent would be 40 watt units, and 30 per cent would be 50 watt units.

21 One of the uncertainties of the GHG emission reductions that may actually be achievable by hydro plants over the sort of timescales appropriate to analysis of climate change is the impact of siltation. Recent surveys of the impoundments in the Upper

Mahaweli scheme show high rates of physical soil loss in the catchment areas, and correspondingly high rates of reservoir siltation. The siltation level of the Polgolla reservoir in 1994, some 14 years after commissioning, has been estimated at 44 per cent of its total capacity. The importance of watershed management has been recognized in Sri Lanka's National Environmental Action Plan, and is one of the main components of a new World Bank supported project to strengthen institutional capacity in watershed management and environmental regulation.

22 However, as noted in Chapter 5, under the assumption of tied concessionary financing (that is, financing that would not be available to Sri Lanka for other sectors), these hydro plants are also in the least (economic) cost plan and would enter the least-cost plan at lower discount rates (for example, in one sensitivity analysis conducted by CEB for a zero discount rate, all these hydro candidates are selected).

23 The least-cost solution could be further improved if we assume additional quantities of residual oil available for diesel-powered generation from a future refinery expansion. However, given the uncertainties of such an expansion, we limit residual oil use to that presently available.

24 However, this is the undiscounted value. When properly discounted (as is necessary for an analysis of carbon offset revenues), the corresponding cost of avoided carbon climbs from US$ 16.7/tonne to US $ 79.2/tonne.

25 The Sri Lanka estimate is that of Electrowatt, as described previously, and includes the cost of the pier and unloading facilities. The Andhra Pradesh costs are described as economic costs (and therefore exclude escalation and IDC), and 'include the cost of pier construction' (and thus are comparable to the Sri Lanka basis). The explanation of the difference is that the Andhra Pradesh costs are based upon the actual proposals of IPPs for MoU-based projects, while the Electrowatt estimate is based upon costs revealed by ICB. However, costs for LNG-combined cycle plants in Andhra Pradesh, as in Sri Lanka, are both based on ICB (in Andhra Pradesh, based upon the cost of the Vijeswaram-II plant).

26 According to ASCI (1999), Andhra Pradesh's own coal resources are limited, and thus can provide only some small share of the production of the large coal fields in the Talcher region of Orissa (which requires shipment by rail to Paradeep, then coastal sea transportation to ports on the Andhra Pradesh coast).

27 A wide range of studies was examined for this exercise. The World Bank's Urban Transport Strategy Review (World Bank, 1996b) presents the growth of vehicle ownership rates (usually measured as vehicles owned per 1000 persons) for France, Spain and Japan. Shoup and Stark (2000) trace the growth of the US vehicle fleet from 1900 to 1995. By tracing the growth of per capita incomes for the same period the relationship between the growth of incomes and vehicle ownership in the US too could be established. In a similar manner, Singapore's vehicle growth as a function of its per capita income has also been traced using data from the Singapore Department of Statistics. The vehicle fleet in the UK was studied using the database at URL: www.Statsbase.gov.uk/statbase/timezone.asp.

10. Dealing with uncertainty: a real options framework for carbon trading*

The results outlined in the previous chapter represent only the starting point for identifying an appropriate GHG strategy for Sri Lanka. Foremost among the problems facing decision-makers is the large uncertainty associated with such calculations. And among these many sources of uncertainty is the question of the value of the carbon offset – which could range from zero if a market for carbon reductions does not develop, to values in excess of US$ 20/tonne if the physical impacts of increasing global warming begin to appear in the short term.

10.1 SOURCES OF UNCERTAINTY AND THE REAL OPTIONS APPROACH

Uncertainty is a key driver that calls for the options-based approach to be used to enhance the traditional NPV rule used in the assessment of sustainable development projects. Jeepma and Munasinghe (1998) classify uncertainty related to climate change into two broad categories:

- Scientific uncertainty, which arises due to limited knowledge of the rates of natural emission and absorption, and of the environmental consequences of greenhouse gases; and
- Socioeconomic and technological uncertainty linked to human actions, including uncertainties which arise due to the difficulty of predicting the future rates of emissions arising from human activity, the inability to value in monetary terms the future impact of greenhouse gas accumulation and other forms of environmental degradation, lack of knowledge about human responses to environmental changes, and lack of knowledge about the development of technological options which would affect all aspects of the environmental equation.

As noted earlier, for example, in the case of dendro-thermal plants there is

* Chapter 10 contributor: Chitru Fernando.

uncertainty about whether fuelwood can actually be grown at the scale and cost envisaged by its advocates, about the cost of gasifier technology, and about the value of carbon emissions avoided from fossil fuels that are displaced.

It is also quite evident that long time frames are very much a factor in sustainable development projects, with typical forecast periods starting in 2010 and ranging up to 100 years or more beyond this time (Nordhaus, 1991; Jeepma and Munasinghe, 1998). Even a relatively small amount of uncertainty can lead to a very wide range of outcomes over such long time periods (Munasinghe and Swart 2005).

Resolving Uncertainties Using the Options Approach

In this section, we follow the model developed in Fernando and Munasinghe (1998). An important first step in the bottom-up approach to mitigation is the identification of the options available for reducing emissions. The standard abatement cost curve for GHG mitigation options is obtained by ranking specific GHG mitigation projects in the order of cost (or benefit in the case of "win–win" projects) of reducing a unit of carbon emissions – as derived in the previous chapter. The cost and benefit estimates of GHG mitigation are obtained by discounting future streams of costs and benefits back to the present time. This ranking implies that projects with a lower cost per unit of carbon reduction should be implemented prior to higher cost projects.

However, this static approach can be somewhat misleading in the recommendations that it provides, especially in the case of GHG mitigation projects which can vary widely with regard to their time frame, the cost of reversal, and the uncertainty associated with future costs and benefits. This is because the standard abatement cost curve (SACC), which is based on the assessment of future costs and benefits at a particular instant in time, disregards the value of the options embedded in specific projects associated with changing the timing of their implementation. Taking proper account of such options could very well lead to a reversal in the ordering of projects along the SACC or control cost staircase.

For example, higher cost (but lower option-valued) projects might be undertaken prior to their lower cost (but higher option-valued) counterparts. This option value will arise, in part, from the fact that the control cost curve is itself dynamic, with changes brought about in the course of time due to the evolution of science and technology, as well as socioeconomic conditions. Developing a dynamic abatement cost curve would require taking account of the significance of each of the above drivers of option value, for each of the GHG mitigation options that have been identified in the standard abatement cost curve. In the following discussion, we first explain the development of the standard abatement cost curve, thereafter turning to how it might change in a dynamic context.[1]

The Dynamic Abatement Cost Curve (DACC)

Uncertainties of various kinds will affect the choice of policy options significantly. First, scientific uncertainties arise from the unpredictability of future climate change and its physical, ecological and socioeconomic impacts. Second, technological uncertainties exist with regard to the evolution of future energy producing and using technology costs and fuel prices. Finally, sociopolitical uncertainties also abound, including the degree of consumer backlash against rising energy prices, the unpredictability of energy demand, the progress in negotiations under the United Nations Framework Convention on Climate Change (UNFCC), the evolution of the carbon market, and so on.

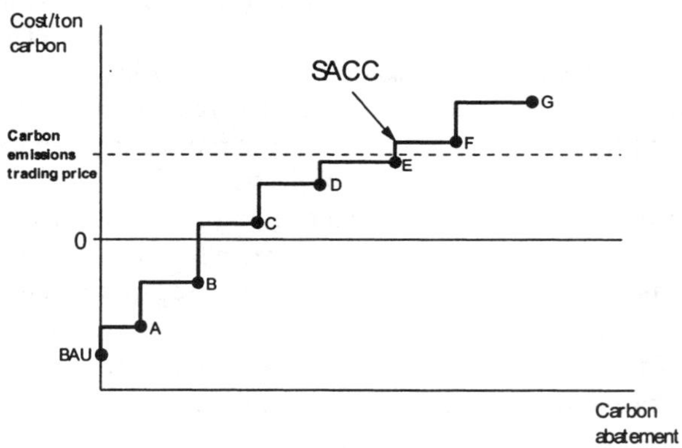

Figure 10.1 The standard abatement cost curve (SACC)

To show how risk and uncertainty might affect decision-making, consider the example shown in Figure 10.1. Let step A represent an increase in energy prices (for example, removal of subsidies to bring prices closer to economically efficient levels), while step B indicates another option such as energy conservation. In a riskless analysis, the decision-maker would first carry out step A, and then proceed to step B. However, in the real world, public reaction against energy price increases may involve too great a political risk, so that it would be more feasible to undertake step B first, and defer step A to a later date. Thus, the introduction of political uncertainty and risk will change the decision-making process.

A second example involves the use of option values. Step D might require commitment to a technology having high and unrecoverable sunk costs. Assuming that step E involves a smaller investment, together with the possibility of switching to another more cost-effective technology (which may become available in the future). In this case, step E has an additional "option" value over step D, because the former preserves the flexibility to change to a potentially cheaper and superior technology (at little cost), in the future. Once again, risk-based option value analysis could well favour step E over step D.

Risk associated with the unpredictability of the global trading price for carbon abatement provides a third example. The future emissions trading price might rise rapidly in the future if climate change effects manifested themselves earlier and more strongly than currently anticipated (or vice versa). An options-based analysis would help to determine the profitability of exercising any specific carbon abatement option on the SACC staircase today, versus retaining the flexibility to use it at a later time.

This decision problem can be analysed using a real options framework, which has been extensively developed in other fields, and which has potential application to GHG mitigation strategy. Box 10.1 illustrates the basic concept using a simple numerical example. A detailed description of the analytical framework is given in Fernando and Munasinghe (1998).

Box 10.1 Calculation of option values

Consider an environmental project, which requires an investment of US$ 100 million. Based on the information available today (time 0), this project will yield an environmental credit valued at US$ 200 million (good state) or US$ 50 million (bad state), with equal probability, a year from now (time 1). The project terminates at time 1. Assuming a discount rate of 10 per cent, the expected value of the investment (NPV) is given by:

$$NPV = \underbrace{0.5\ \{200/1.1\}}_{\substack{\text{return in year 1} \\ \text{x Pr\{good state\}}}} + \underbrace{0.5\ \{50/1.1\}}_{\substack{\text{return in year 1} \\ \text{x Pr\{bad state\}}}} - \underbrace{100}_{\substack{\text{cost of} \\ \text{investment}}} = US\$\ 13.6\ million$$

The standard NPV rule would recommend that this project be undertaken in order to realize the expected positive net present value of US$ 13.6 million.

Now suppose the project can be postponed at a cost of US$ 5 million. Suppose further that one year from now it will be known whether the state is good or bad. If the good state is observed, the investment will be made, with a return of US$ 200 million at time 2. If the bad state is observed, the investment will not be made. Therefore the NPV of postponing the investment decision until $t = 1$ is given by

$$NPV = \underbrace{0.5\ \{200/1.1^2\}}_{\substack{\text{Return in year 2} \\ \text{x Pr\{good state\}}}} - \underbrace{0.5\{100/1.1\}}_{\substack{\text{cost of} \\ \text{investment}}} + \underbrace{0.5 * zero}_{\substack{\text{Pr\{bad state\}} \\ \text{x [no investment]}}} - \underbrace{5}_{\substack{\text{cost of} \\ \text{delay}}} = US\$\ 32.2 million$$

Thus, despite having a positive NPV if undertaken at time 0, it will never be optimal to invest in this project at time 0, since investors can always do better by waiting until time 1, when the uncertainty is resolved, and the expected value increases to US$ 32 million. The option value associated with delaying a decision by one year is therefore 32.2-13.6 = US$18.6 million.

10.2 APPLYING THE OPTIONS APPROACH

We begin the analysis of uncertainty in Table 10.1, in which we examine the use of oil steam (+FGD) as a mitigation measure, given the least-cost scenario as a starting point. As shown in panel A, this increases the system cost by US$9.7 million and decreases GHG emissions by 8.1 million tonnes, with a discounted cost of avoided carbon of US$21.2/tonne (which falls to US$4.4/tonne when undiscounted).[2]

Table 10.1 Oil-steam (+FGD) compared to least-cost solution

A. Cost and avoided carbon if implemented today

			Baseline: LeastCost	Mitigation measure ++OilSt+FGD	Delta
[1]	System cost	[$USmill.]	2392.6	2402.3	9.7
	Undiscounted				
[2]	GHG[CO2]	[mill.tons]	87.8	79.6	-8.1
[3]	Cost of avoided CO2	[$/tCO2]			1.2
[4]	Cost of avoided carbon	[$/tC]			4.4
	Discounted				
[5]	GHG[CO2]	[mill.tons]	25.4	23.8	-1.7
[6]	Cost of avoided CO2	[$/tCO2]			5.8
[7]	Cost of avoided carbon	[$/tC]			21.2

B. Expected value of revenue from carbon offsets

		Pr{ }	R(c) [$million]	E{R(c)} [$million]
[1]	$0/ton	0.025	0.00	0.0
[2]	$10/ton	0.150	4.58	0.7
[3]	$20/ton	0.500	9.15	4.6
[4]	$30/ton	0.250	13.73	3.4
[5]	$40/ton	0.075	18.31	1.4
[6]	Discounted value			10.1

C: Expected value of implemeting measure today ($US million)

[1]	Cost with mitigation measure (from A, line[1])	++OilSt+FGD	2402.3
[2]	Less expected value of carbon offsets (from B,line[8])		-10.1
[3]	Expected cost of mitigation measure =[1]+[2]		2392.3
[4]	Cost without mitigation measure (from A,line[1])	LeastCost	2392.6
[5]	Hence cost reduction		0.4
[6]			benefit: proceed

D. Cost and avoided carbon if measure :++OilSt+FGD is implemented at t+5

			Baseline: Reform+5	Mitigation measure ++Dendro	Delta
	+Interim measure:=		diesel	diesel	
[1]	System cost	[$USmill.]	2434	2441	7.1
	Undiscounted				
[2]	GHG[CO2]	[mill.tons]	84.7	78.5	-6.3
[3]	Cost of avoided CO2	[$/tCO2]			1.1
[4]	Cost of avoided carbon	[$/tC]			4.1
	Discounted				
[5]	GHG[CO2]	[mill.tons]	24.6	23.5	-1.2
[6]	Cost of avoided CO2	[$/tCO2]			6.0
[7]	Cost of avoided carbon	[$/tC]			21.9

E. Impact of delay

			Baseline: Reform+5	Mitigation measure ++Dendro
[1]	system cost	[$USmillion]	41.6	39.0
[2]	GHG[CO2]undiscounte	[mill.tons]	-3.1	-1.2

Table 10.1 (continued)

F. Decision if uncertainty resolved: calculation of option value

	Pr{ }	PV[E{R(c)}]	++Dendro	Reform+5	Delta Cost	Decision	Benefit	E{benefit}
[1] $0/ton	0.025	0.0	2441	2434	7	do not build	0.0	0.0
[2] $10/ton	0.150	3.2	2438	2434	4	do not build	0.0	0.0
[3] $20/ton	0.500	6.5	2435	2434	1	do not build	0.0	0.0
[4] $30/ton	0.250	9.7	2432	2434	-3	build	2.6	0.7
[5] $40/ton	0.075	12.9	2428	2434	-6	build	5.8	0.4
[6] Total, expected value of carbon offsets								1.1
[7] Benefit if implemented today (from C, line [5])								0.4
[8] Hence option value of delay ([6]-[7])								0.7

Panel B shows the probability distribution for the value of the carbon offset, expressed as US$/tonne Carbon. The expected value of the carbon offset, E{V(c)}, is US$ 22.00. The corresponding expected value of carbon offset revenue, E{R(c)}, is US$ 10.1 million when discounted at 10 per cent. E{R(c)} is given by :

$$E\{R(c)\} = \Sigma\ VjP\{j\}\Delta C$$

where ΔC is the avoided carbon (8.1 million tonnes in this case), and P{j} is the probability of the carbon offset having value Vj.

Panel C then derives the expected value of implementing this mitigation option today, which is the cost as given in Table 9.20 (NPV of US$ 2,402.3 million), less the expected value of the revenue from carbon offsets (US$ 10.1 million). This is US$ 0.4 million less than the least-cost case (US$ 2,392.6 million) whose expansion plan is depicted in Figure 9.14. This reduction in cost (though quite small – about 4 per cent of the expected value of carbon offset revenue) would normally justify undertaking the mitigation measure (given the expected value of US$ 22/tonne for carbon offsets).[3]

In Panel D we now consider the implications of delaying a decision by five years. Since a commitment to coal baseload plant in the least-cost solution would otherwise need to be made now (to meet demands five years hence), this delay is not costless, for in its place one must build some other, interim capacity, say, diesels, to meet the demand. We therefore recalculate the capacity expansion plan such that the first coal unit is delayed by five years, with 300 MW of diesels (at US$ 1,100/kW) built over the five-year period. Then five years from now, one would reconsider whether to then build a coal plant, or whether to implement the mitigation measure. The new costs and GHG emissions are noted in Panel D. Because the presence of the interim diesels has some impact on the merit order dispatch, GHG emissions are different than those shown in Panel A, with a slightly higher cost of avoided carbon (US$ 21.9/tonne, row[6] of panel D).

In panel E we show the impacts of delay. The costs increase in both cases by a roughly similar amount.[4] We know that the least-cost case builds a first coal plant in 2006: it necessarily follows that if one were to delay the coal plant and build something else to meet the demand in the interim, costs will increase – by US$41.6 million. Greenhouse gas emissions, however, decrease by 3.1 million tonnes, because emissions from a diesel plant are lower than from the corresponding generation from coal.

Similarly, because diesels have slightly greater efficiency than steam-cycle plants, replacing the first oil plant by diesels also reduces GHG emissions relative to no delay, and so GHG emissions decrease by 1.2 million tonnes per year.

In panel F we now recalculate the expected values given resolution of the uncertainty at T + 5. If the actual value of the offset is less than the cost of avoided carbon (US$ 21.9/tonne), then the mitigation measure is not undertaken; but it is undertaken if the actual value is greater than US$ 21.9/tonne – which in the case of the hypothetical discrete probability distribution assumed here is the case for actual offset values of US$ 30/tonne and US$ 40/tonne. The expected value of benefits is calculated at US$ 1.09 million, which is greater than the US$ 0.37 million calculated if the decision is made today. In other words the option value is US$ 1.09 - 0.37 = US$ 0.72 million.

While the numbers are small, nevertheless we note that the option value of delay is almost twice the expected value of the benefit of making the commitment to the mitigation measure today – and represents about 7 per cent of the expected value of carbon offset revenue.

However, perhaps more important is the illustration of the importance of proper discounting. The GEF suggests that carbon savings be expressed in undiscounted terms. Since undiscounted carbon emissions over some given planning horizon will always be greater than the corresponding discounted sum, it necessarily follows that the apparent cost of avoided carbon will be lower than if discounted. However, the economic flows that result from carbon offset payments must be discounted for valid analysis, which effectively increases the cost of avoided carbon (from US$4.4/tonne to US$21.2/tonne in this example).[5]

The option value will vary with the extent of uncertainty: the higher the variance of the probability distribution for the future value of the carbon offset, the greater the option value (Figure 10.2).

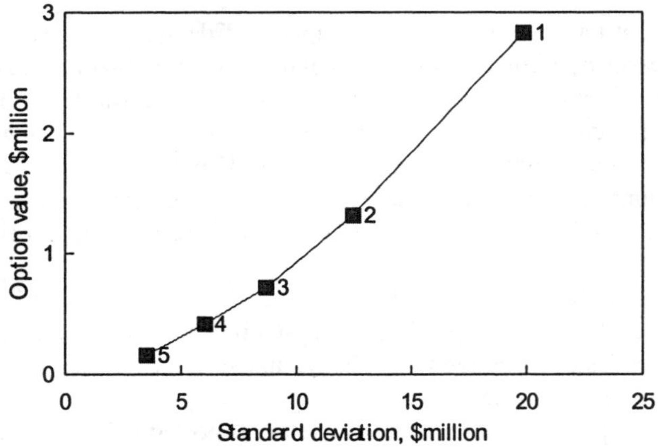

Figure 10.2 Option value as a function of the standard deviation of the probability distribution of the carbon offset payment

Figure 10.3 shows the details of this calculation. It also shows the option value, again for oil-steam+ FGD, for five different probability distributions, each with expected value of US$22/tonne, but with different variances. The probability distribution used in Table 10.1 corresponds to probability distribution #3 in this figure (with standard deviation of US$7.8 million). We note that in the case of high uncertainty (as in probability distribution #1), the option value increases to US$2.85 million, some seven times greater than the expected value of a decision to proceed today.

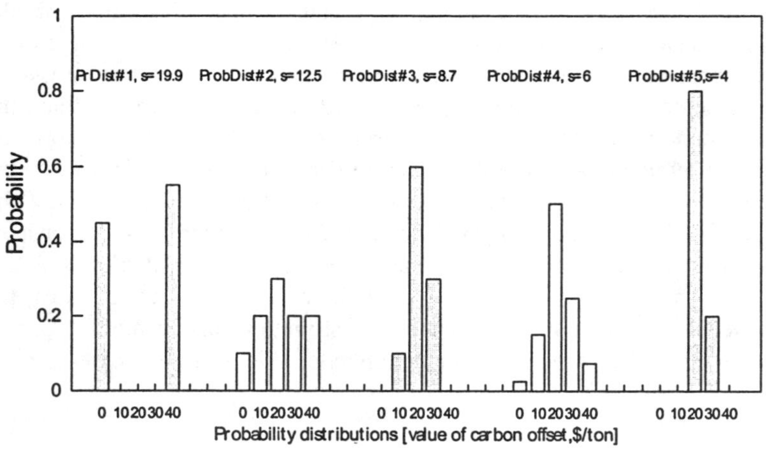

Figure 10.3 Probability distributions

Table 10.2 shows the results for LNG (measured against the reform option). This shows that a commitment to LNG today has an expected value of minus US$ 69.5 million: in other words, the costs of the carbon offsets are insufficient to compensate for the higher costs. This follows logically given a cost of avoided carbon of US$ 79.2/tonne compared to an expected carbon offset value of US$ 22/tonne.

Table 10.2 LNG compared to reform

A. Cost and avoided carbon if implemented today

		Baseline: Reform	Mitigation measure LNG	Delta
[1] System cost	[$USmill.]	2448.8	2545.1	96.3
Undiscounted				
[2] GHG[CO2]	[mill.tons]	98.4	77.3	-21.1
[3] Cost of avoided CO2	[$/tCO2]			4.6
[4] Cost of avoided carbon	[$/tC]			16.7
Discounted				
[5] GHG[CO2]	[mill.tons]	28.2	23.8	-4.5
[6] Cost of avoided CO2	[$/tCO2]			21.6
[7] Cost of avoided carbon	[$/tC]			79.2

B. Expected value of revenue from carbon offsets

	Pr{ }	R(c) [$million]	E{R(c)} [$million]
[1] $0/ton	0.025	0.00	0.0
[2] $10/ton	0.150	12.15	1.8
[3] $20/ton	0.500	24.30	12.1
[4] $30/ton	0.250	36.45	9.1
[5] $40/ton	0.075	48.60	3.6
[6] Discounted value			26.7

C: Expected value of implemeting measure today ($US million)

[1] Cost with mitigation measure (from A, line[1])	LNG	2545.1
[2] Less expected value of carbon offsets (from B,line[8])		-26.7
[3] Expected cost of mitigation measure =[1]+[2]		2518.4
[4] Cost without mitigation measure (from A,line[1])	Reform	2448.8
[5] Hence cost reduction		-69.5
[6]		cost: do not proceed

D. Cost and avoided carbon if measure :LNG is implemented at t+5

+Interim measure:=		Baseline: Reform+5 diesel	Mitigation measure ++Dendro diesel	Delta
[1] System cost	[$USmill.]	2515	2566	50.7
Undiscounted				
[2] GHG[CO2]	[mill.tons]	94.3	79.5	-14.8
[3] Cost of avoided CO2	[$/tCO2]			3.4
[4] Cost of avoided carbon	[$/tC]			12.6
Discounted				
[5] GHG[CO2]	[mill.tons]	27.1	24.3	-2.8
[6] Cost of avoided CO2	[$/tCO2]			18.2
[7] Cost of avoided carbon	[$/tC]			66.6

Table 10.2 (continued)

E. Impact of delay

		Baseline: Reform+5	Mitigation measure ++Dendro
[1] system cost	[$USmillion]	66.2	20.6
[2] GHG[CO2]undiscounte	[mill.tons]	-4.1	2.3

F. Decision if uncertainty resolved: calculation of option value

	Pr{ }	PV[E{R(c)}]	++Dendro	Reform+5	Delta Cost	Decision	Benefit	E{benefit}
[1] $0/ton	0.025	0.0	2566	2515	51	do not build	0.0	0.0
[2] $10/ton	0.150	7.6	2558	2515	43	do not build	0.0	0.0
[3] $20/ton	0.500	15.2	2551	2515	35	do not build	0.0	0.0
[4] $30/ton	0.250	22.8	2543	2515	28	do not build	0.0	0.0
[5] $40/ton	0.075	30.4	2535	2515	20	do not build	0.0	0.0
[6] Total, expected value of carbon offsets								0.0
[7] Benefit if implemented today (from C, line [5])							cost!	-69.5
[8] Hence option value of delay ([6]-[7])								no option value!

Waiting for five years does not change the attractiveness of LNG, and there is no option value. Obviously, even if the uncertainty is resolved after five years, if the avoided cost of carbon is US$ 66/tonne (Panel D, row [6]), then even if the actual value of the carbon offset is US$ 40/tonne, the LNG option would remain unattractive.

Table 10.3 shows the option value calculations for the dendro-thermal option. The expected value of a commitment today is a benefit of US$ 31.4 million, which again follows immediately from a comparison of the cost of avoided carbon (US$ 10.9/tonne) with the expected value of the carbon offset.

However, the impact of delay is quite different to the two cases examined above. As before, forcing diesels into the least-cost solution for the interim period reduces GHG emissions. But in the case of the dendro-thermal option, diesels will significantly increase GHG emissions – by 9.1 million tonnes (Panel E, row [2]). This results in an aggregate cost of avoided carbon that jumps to US$ 42.6/tonne (Panel D, row[6]). Hence given the maximum value of US$ 40/tonne in the discrete probability distribution assumed for carbon offset revenue, dendro-thermal would not be implemented at t+5. Clearly, there is no option value in this case.

This counter-intuitive result is readily explained: since the diesels have an economic life that extend considerably beyond the assumed five-year interim period, new dendro-thermal plants at t+5 must now compete only against the variable cost of diesel plant operation, since the capital costs are sunk, and no longer relevant at t+5.

Table 10.3 Dendro-thermal relative to least-cost solution

A. Cost and avoided carbon if implemented today

			Baseline: Reform	Mitigation measure ++Dendro	Delta
[1]	System cost	[$USmill.]	2448.8	2482.4	33.6
	Undiscounted				
[2]	GHG[CO2]	[mill.tons]	98.4	44.9	-53.5
[3]	Cost of avoided CO2	[$/tCO2]			0.6
[4]	Cost of avoided carbon	[$/tC]			2.3
	Discounted				
[5]	GHG[CO2]	[mill.tons]	28.2	16.9	-11.3
[6]	Cost of avoided CO2	[$/tCO2]			3.0
[7]	Cost of avoided carbon	[$/tC]			10.9

B. Expected value of revenue from carbon offsets

		Pr{ }	R(c) [$million]	E{R(c)} [$million]
[1]	$0/ton	0.025	0.00	0.0
[2]	$10/ton	0.150	29.53	4.4
[3]	$20/ton	0.500	59.07	29.5
[4]	$30/ton	0.250	88.60	22.2
[5]	$40/ton	0.075	118.14	8.9
[6]	Discounted value			65.0

C: Expected value of implemeting measure today ($US million)

[1]	Cost with mitigation measure (from A, line[1])	++Dendro	2482.4
[2]	Less expected value of carbon offsets (from B,line[8])		-65.0
[3]	Expected cost of mitigation measure =[1]+[2]		2417.4
[4]	Cost without mitigation measure (from A,line[1])	Reform	2448.8
[5]	Hence cost reduction		31.4
[6]			benefit: proceed

D. Cost and avoided carbon if measure :++Dendro is implemented at t+5

			Baseline: Reform+5	Mitigation measure ++Dendro	Delta
	+Interim measure:=		dendro demo	demo	
[1]	System cost	[$USmill.]	2515	2482	-32.6
	Undiscounted				
[2]	GHG[CO2]	[mill.tons]	94.3	44.9	-49.5
[3]	Cost of avoided CO2	[$/tCO2]			-0.7
[4]	Cost of avoided carbon	[$/tC]			-2.4
	Discounted				
[5]	GHG[CO2]	[mill.tons]	27.1	16.9	-10.2
[6]	Cost of avoided CO2	[$/tCO2]			-3.2
[7]	Cost of avoided carbon	[$/tC]			-11.7

E. Impact of delay

			Baseline: Reform+5	Mitigation measure ++Dendro
[1]	system cost	[$USmillion]	66.2	0.0
[2]	GHG[CO2]undiscounte	[mill.tons]	-4.1	0.0

Table 10.3 (continued)

	Pr{ }	PV[E{R(c)}]	++Dendro	Reform+5	Delta Cost	Decision	Benefit	E{benefit}
[1] $0/ton	0.025	0.0	2602	2515	87	do not build	0.0	0.0
[2] $10/ton	0.150	20.4	2582	2515	67	do not build	0.0	0.0
[3] $20/ton	0.500	40.8	2561	2515	46	do not build	0.0	0.0
[4] $30/ton	0.250	61.2	2541	2515	26	do not build	0.0	0.0
[5] $40/ton	0.075	81.6	2520	2515	5	do not build	0.0	0.0
[6] Total, expected value of carbon offsets								0.0
[7] Benefit if implemented today (from C, line [5])								31.4
[8] Hence option value of delay ([6]-[7])								no option value!

However, one of the advantages of dendro-thermal is that it can be implemented in small capacity increments. Unlike a commitment to a coal plant, which would be for at least 600–900 MW to capture the necessary economies of scale, dendro-thermal can be built in increments of as little as 1–5 MW. Therefore, instead of the intermediate five-year commitment to diesels before choosing coal or dendro-thermal, we make a five-year commitment to a dendro-thermal demonstration programme, and then decide whether to go to coal, or continue with a full-scale dendro-thermal programme.

The corresponding option value calculation is presented in Table 10.4. The expected value for a present permanent commitment to dendro-thermal is unchanged at US$ 31.4 million. However, with a dendro-thermal demonstration as the interim measure, the impact of delay is to create a win–win situation, for the mitigation measure has costs of US$ 2,482 million that are below that of the least-cost solution (again with the interim dendro-thermal demonstration) of US$ 2,515 million, though of course still higher than the least-cost solution alone (of US$ 2,448 million). However, the calculated option value of US$ 59.4 million indicates the significant value of being able to abandon the dendro-thermal option at t + 5, regardless of the value of the carbon offset.

Table 10.4 Dendro-thermal option relative to least-cost solution, with dendro-thermal demonstration as the interim option

A. Cost and avoided carbon if implemented today

			Baseline: Reform	Mitigation measure ++Dendro	Delta
[1]	System cost	[$USmill.]	2448.8	2482.4	33.6
	Undiscounted				
[2]	GHG[CO2]	[mill.tons]	98.4	44.9	-53.5
[3]	Cost of avoided CO2	[$/tCO2]			0.6
[4]	Cost of avoided carbon	[$/tC]			2.3
	Discounted				
[5]	GHG[CO2]	[mill.tons]	28.2	16.9	-11.3
[6]	Cost of avoided CO2	[$/tCO2]			3.0
[7]	Cost of avoided carbon	[$/tC]			10.9

B. Expected value of revenue from carbon offsets

		Pr{ }	R(c) [$million]	E{R(c)} [$million]
[1]	$0/ton	0.025	0.00	0.0
[2]	$10/ton	0.150	29.53	4.4
[3]	$20/ton	0.500	59.07	29.5
[4]	$30/ton	0.250	88.60	22.2
[5]	$40/ton	0.075	118.14	8.9
[8]	Discounted value			65.0

C: Expected value of implemeting measure today ($US million)

[1]	Cost with mitigation measure (from A, line[1])	++Dendro	2482.4
[2]	Less expected value of carbon offsets (from B,line[8])		-65.0
[3]	Expected cost of mitigation measure =[1]+[2]		2417.4
[4]	Cost without mitigation measure (from A,line[1])	Reform	2448.8
[5]	Hence cost reduction		31.4
[6]			benefit: proceed

D. Cost and avoided carbon if measure :++Dendro is implemented at t+5

			Baseline: Reform+5	Mitigation measure ++Dendro	Delta
	+Interim measure:=		dendro demo	demo	
[1]	System cost	[$USmill.]	2515	2482	-32.6
	Undiscounted				
[2]	GHG[CO2]	[mill.tons]	94.3	44.9	-49.5
[3]	Cost of avoided CO2	[$/tCO2]			-0.7
[4]	Cost of avoided carbon	[$/tC]			-2.4
	Discounted				
[5]	GHG[CO2]	[mill.tons]	27.1	16.9	-10.2
[6]	Cost of avoided CO2	[$/tCO2]			-3.2
[7]	Cost of avoided carbon	[$/tC]			-11.7

E. Impact of delay

			Baseline: Reform+5	Mitigation measure ++Dendro
[1]	system cost	[$USmillion]	66.2	0.0
[2]	GHG[CO2]undiscounte	[mill.tons]	-4.1	0.0

Table 10.4 (continued)

F. Decision if uncertainty resolved: calculation of option value

	Pr{ }	PV[E{R(c)}]	++Dendro	Reform+5	Delta Cost	Decision	Benefit	E{benefit}
[1] $0/ton	0.025	0.0	2482	2515	-33	build	32.6	0.8
[2] $10/ton	0.150	26.4	2456	2515	-59	build	59.1	8.9
[3] $20/ton	0.500	52.9	2430	2515	-86	build	85.5	42.8
[4] $30/ton	0.250	79.3	2403	2515	-112	build	112.0	28.0
[5] $40/ton	0.075	105.8	2377	2515	-138	build	138.4	10.4
[6] Total, expected value of carbon offsets								90.8
[7] Benefit if implemented today (from C, line [5])								31.4
[8] Hence option value of delay ([6]-[7])								59.4

NOTES

1 This discussion is based on a more detailed paper, Fernando and Munasinghe (1998).

2 The figures in panel A for system cost, GHG emissions and avoided cost match the data given in Table 9.18.

3 We would of course expect this mitigation measure to be undertaken, since the cost of avoided carbon (US$ 21.2/tonne) is below the expected value of the carbon offset revenue (at US$ 22.0/tonne)!

4 Even though 300 MW of diesels are built in each case, the increase in cost is not identical because of the varying impact on subsequent merit order dispatch (and fuel costs). The calculations of system cost and GHG emissions are again done in ENVIROPLAN; the interim measure – in this case, diesels – must necessarily be forced in (since diesels are not part of the least-cost solution).

5 At the same time, one must be sure that end-effects are properly accounted for, since the life of some mitigation measures will extend beyond the planning horizon. ENVIROPLAN accounts for these end-effects by calculating a salvage value of assets whose economic life extends beyond the planning horizon, and whose value is subtracted from the system cost.

11. Renewable energy for rural electrification

This chapter discusses the economic analysis of renewable energy for rural electrification. Rural electrification has important implications for poverty reduction. As in most Asian countries, poverty is largely a rural phenomenon: a 1996 survey concluded that 88 of the poor live in rural areas, compared to 8 per cent in urban areas and 4 per cent in estate areas (Nexant, 2004). At the end of 2002, only 63 per cent of all households in Sri Lanka had electricity with electrification percentages ranging from 87 per cent in Colombo to 32 per cent in Moneragala. Indeed, access to electricity and poverty are highly correlated, as shown in Figure 11.1: the higher the poverty rate, the lower the electrification rate. The data shown do not include districts in the northern and eastern areas most affected by ethnic conflict, where poverty rates are substantially higher, and electrification rates lower.

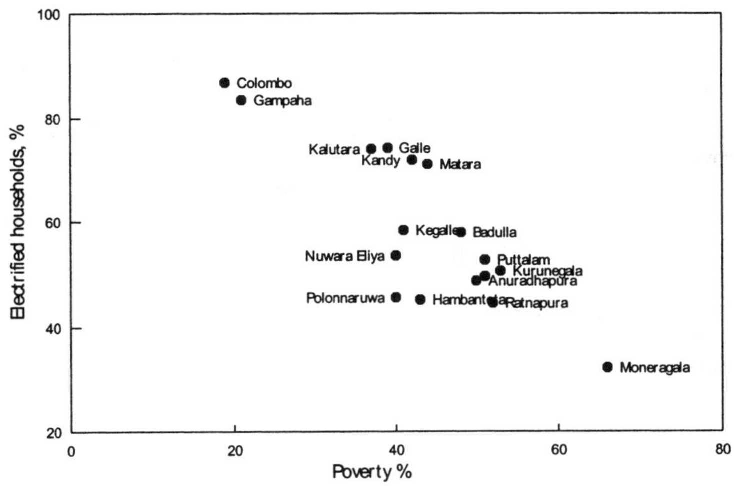

Source: Based on data in Nexant (2004) and Sri Lanka Central Bank

Figure 11.1 Electrification and poverty for selected districts

Off-grid schemes account for a small but growing segment of household electrifications: some 28,000 households now have PV solar home systems SHS), and 3,000 households are supplied by village-scale hydro systems. These off-grid options are the subject of this chapter: they represent two of the components of the World Bank/GEF Energy Services Delivery (ESD) programme (the other components supported the 3 MW wind demonstration programme and grid-connected small hydro). However, we begin the chapter with a discussion of one of the main issues, namely how the benefits of rural electrification are to be quantified for economic analysis.

11.1 MEASURING ECONOMIC BENEFITS OF RURAL ELECTRIFICATION

There are two basic methods for estimating economic benefits (see for example Munasinghe 1987). The first is to set benefits equal to the avoided costs of the various devices that are presently used to supply energy, and that would be replaced by the mini-hydro scheme: these include kerosene, diesel generation for auto battery charging and for productive uses (as in rice husking), candles and dry cells. This "avoided cost" method is easily applied because the information required can be provided by surveys of existing households, and the resulting calculations are straightforward. However, for two reasons this method generally underestimates the actual benefits.

The first is that the quality of energy supply from mini-hydro based electricity is far superior to that from most alternative devices: for example, the illumination derived from a compact fluorescent lamp (CFL) is much higher than that provided by candles or kerosene lamps, in addition to avoiding many of the harmful side-effects such as smoke and odour, and the risk of fire and injury). Indeed, in Sri Lanka it is estimated that 50 percent of burn injuries are attributable to accidents involving kerosene lanterns and candles, of which a substantial fraction occurs to children.[1] Similar problems are reported in many other countries (Foster and Tre, 2000). In the Philippines, numerous reports of deaths have been reported from fires caused by kerosene lamps (Reyes, 2001).

This is an important area for further research, because burn injuries and child deaths have high social costs whose avoidance should be counted as one of the benefits of rural electrification in general (and solar PV systems in particular). Indeed, such direct costs may be easier to ascertain, and most likely of greater magnitude, than those related to local air pollution.

Individuals are prepared to pay more for high quality service, which is to say that they value a given number of lumens from hydro-based electricity generation much more than the equivalent number of lumens from candles and kerosene. In other words, the benefits are greater than those that may be inferred from replacement costs alone.

The second is that the demand curve for electricity (or the equivalent service, such as lighting or TV viewing) is not flat. It is well established that individuals are prepared to pay very high prices for the first few kWh of electricity (or lumens), sufficient, say, for one or two CFLs. They are also prepared to pay high prices for enough electricity to power a TV. But the amount they are prepared to pay, say, for the 10th and 11th CFL will be much less than the first and second CFL. This demand curve – the representation of quantity demanded as a function of price – is therefore downward sloping, and the total benefits from some level of consumption is given by the area under the demand curve (to that point).

Thus, the second approach to calculating benefits requires estimation of such a demand curve. This is generally more difficult, because there may be few actual data points available to accurately determine the shape of the curve. But despite the additional uncertainty, this method of estimating the demand curve and the formalization of willingness-to-pay (WTP) is generally accepted as a better measure of the benefits of electrification.

Consider Figure 11.2, which depicts the demand for services (lighting, TV viewing) of a rural household. The demand curve is downward sloping, and typically has a shape that is concave (with respect to the origin). This shape frequently emerges where more than two points on the curve can be plotted (in the figure only the points x and y are assumed known: a concave shape also follows from the (convenient) assumption of constant elasticity (an assumption often made in econometric models).[2]

Before electrification, these services are typically provided by a mix of kerosene (for lighting) and batteries (for TVs and radios). For simplicity, assume that the only service provided is lighting from kerosene lamps: the quantity of services so consumed is Q_{KERO}, at the price P_{KERO}. The total household expenditure on lighting is therefore $Q_{KERO} \times P_{KERO}$, equal to the area B + D.

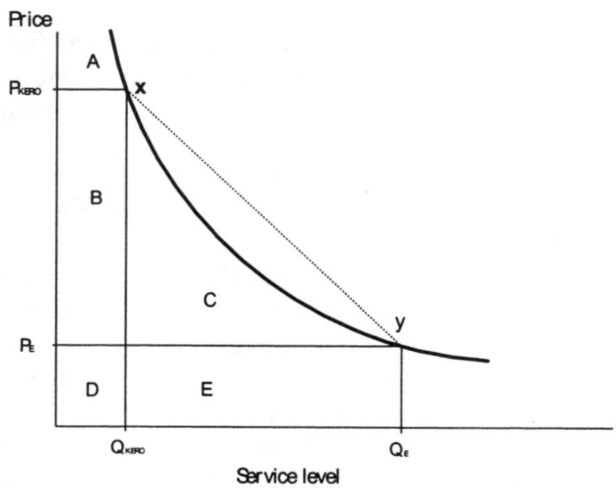

Figure 11.2 Demand curve for electricity provided services

The total willingness to pay (WTP) for the service at level Q_{KERO} is the total area under the demand curve to that level of consumption, that is, areas A + B + D. This is the total benefit to the consumer. However, the cost is area B + D, and therefore the net benefit, also called the consumer surplus, is the difference between the two, namely the area A.

After electrification, the level of service (in the case of lighting, the number of lumen-hours) typically increases substantially; consumption increases from Q_{KERO} to Q_E, but the price paid for the electrified service also falls (typically) from P_{KERO} to P_E. Now the household's expenditure for electricity is $P_E \times Q_E$, equal to the area D+E.

At this level of consumption, the total area under the demand curve to Q_E, that is, the total benefit, is now the area A + B + C + D + E. Therefore the net benefit, or consumer surplus, after subtracting the cost D + E, is A + B + C. Thus, it follows that the net economic benefit of electrification is the increase in consumer surplus, which is the area B + C.[3]

Areas B, D and E are readily calculated from knowledge of consumption before and after electrification, from the household budget for kerosene (and battery charging), and from the tariff of electrified service: that is, given knowledge of the two points on the demand curve x and y, the areas B, D and E are immediately calculable. But area C is more difficult to estimate, since it requires knowledge of the shape of the demand curve between points x and y. The most convenient assumption – and therefore the one most frequently encountered – is that the demand curve is linear.[4]

Unfortunately such an assumption will lead to an over-estimate of the area C, and of the net benefits of electrification, because, as noted, the empirical evidence is that the demand curve is much more likely to have a concave shape, as shown in Figure 11.2: given some functional form for such a demand curve, the area C is readily calculated as the definite integral (see Annex to Chapter 7). The consumer surplus calculations use financial costs (since these are the costs actually seen by the consumer). However, the financial and economic costs require careful enumeration and reconciliation, because often the quantities both before and after electrification are subject to both taxes and subsidies. Figure 11.3 illustrates a typical situation where the price of kerosene is subsidized, and where electrification involves both taxes (for example, VAT on construction costs) and subsidies (as might be provided, for example, by GEF).

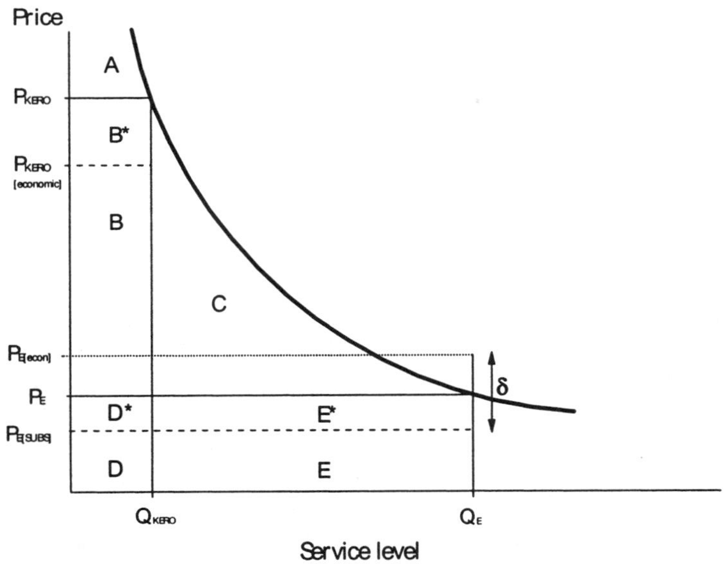

Figure 11.3 Reconciliation of economic and financial flows

Suppose the electrification subsidy, expressed as Rs per kWh, is δ. Then one needs to distinguish between the economic and subsidized prices $P_{E[econ]}$ and $P_{E[subs]}$ respectively: the total amount of the subsidy is the area δ Q_E. Suppose that this subsidized price is also taxed at the rate τ, the price to the consumer then increases to $P_E = P_{E[SUBS]} + \tau$ (the amount of taxes collected being the area D* + E*). As drawn, the assumption is that the subsidy exceeds related taxes (which is the case for the micro-hydro systems, where the proposed subsidy far exceeds the amount of VAT and profits tax on construction). For kerosene, the total (financial) cost to the consumer is B* + B + D* + D. Of this, the area B* represents taxes.

The various economic and financial quantities therefore reconcile as shown in Table 11.1. For example, before electrification, the actual economic benefit is A + B*, but the area B* is captured by government as tax, rather than accruing to the consumer.

Table 11.1 Reconciliation of economic and financial flows

	Kerosene	Electrification	Difference
Consumer surplus	A	A + B* + B + C	B + B* + C
Taxes and duties (VAT)	B*	D* + E*	D* + E* - B*
Subsidy		$-\delta Q_E$	$-\delta Q_E$
Economic cost	B + D* + D	D + E + dQ_E	+ E + dQ_E - B - D*
Total benefit to consumer (= area under demand curve)	A + B* +B + D* + D	A + B + B* + C + D* + D + E* + E	C + E* + E

11.2 THE SOLAR HOMES PROGRAMME

Solar home systems (SHS) are designed to provide the most high-valued, initial tranche of electricity consumption in rural areas that are unlikely to be connected to the grid in the foreseeable future. Nevertheless, according to Shell Solar, the leading supplier of SHS, 10,000 of 17,000 systems sold to date have been to households where the grid was extended after the SHS had been installed. Indeed, there are also reports of households *already* connected to the grid buying a SHS. Evidently the erratic nature of rural supply (2002 was a year of serious power cuts), and CEB's high consumer tariff are powerful incentives for SHS purchases. The solar homes programme received major impetus from the Energy Services Delivery (ESD) project supported by the World Bank and GEF, which have assisted financing of SHS.

The solar industry was at a nascent stage when the ESD project became effective, with only three small operations (Solar Power and Light, Sarvodaya, RESCO) selling roughly 20–30 systems per month in 1998. The ESD project has catalysed the market for solar home systems:

by 2002 sales reached 850 systems per month, achieved by four companies (Shell Solar, Access Solar, SELCO and Alpha Thermal). Some 21,000 systems were installed under the project, and 2003 sales exceeded 10,000 systems.

The very success of the ESD project in stimulating a successful private sector involvement poses difficulties for an economic analysis, because much of the data on actual system costs and margins is confidential. In addition, different companies have also negotiated different deals on tax concessions. However both the strong consumer interest in SHS, and vendors' interest in this market, are testimony to the economic benefits of SHS (given the demonstrated willingness to pay for lighting and TV service at a cost significantly above the cost of kerosene and battery charging), and to the financial benefits (given the financial success of the vendors).

Several surveys provide information on monthly expenditures of households that purchased SHS (Table 11.2). Although the results show considerable variation, the data do indicate a substantial willingness-to-pay for the safety, convenience and superior light quality that is obtainable from a SHS.[5]

Table 11.2 Summary of survey results

	Masse (2001)	IRG (2002)	AC Nielson (2003)
Monthly kerosene expenditure, Rs	231	340	312
Total monthly energy expenditure before SHS	321	582	495
Average monthly energy expenditure after SHS	–	1,450	1,286
Sample size	–	100	250

Figure 11.4 shows monthly energy expenditure of 98 sample households on kerosene, dry cell batteries, and rechargeable batteries (used for TVs), sorted in ascending order, prior to acquisition of the SHS.[5] The monthly averages are Rs 340 for kerosene, Rs 92 for dry cells, and Rs 152 for battery charging for a total of Rs 582 per month.

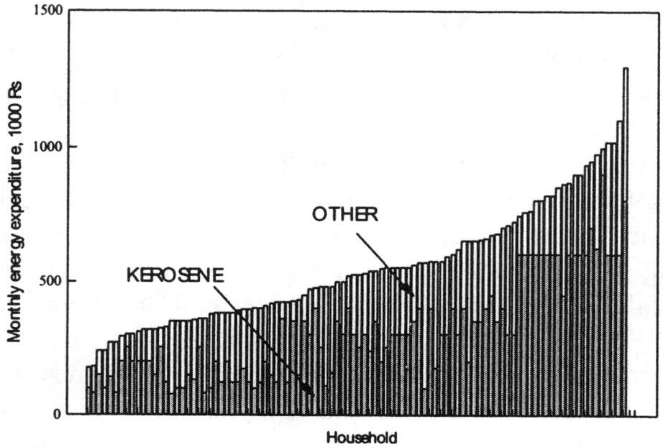

Figure 11.4 Household energy expenditures, before SHS

However, we note little correlation between energy expenditure and reported disposable income (among households that subsequently bought SHS), as shown in Figure 11.5.

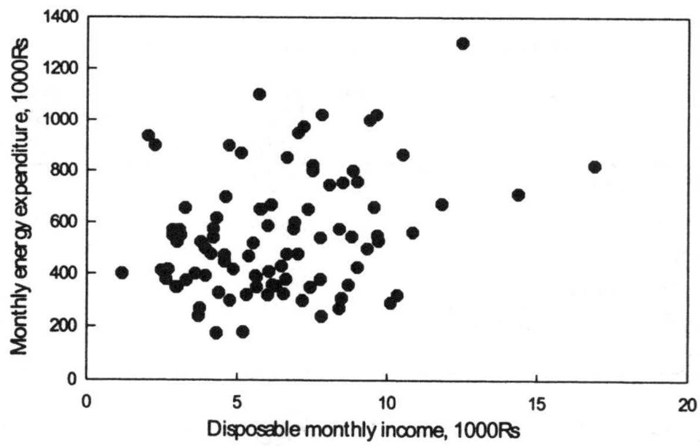

Figure 11.5 Energy expenditure v. income (before SHS)

The monthly energy expenditures increased sharply upon purchase of the SHS (see Figure 11.6). Households recognized, however, that these sharply higher expenditures would be of limited duration during the debt repayment period, and that thereafter their energy expenditures would decrease sharply. The average monthly energy expenditure of the sample households with SHS was Rs 1,450.

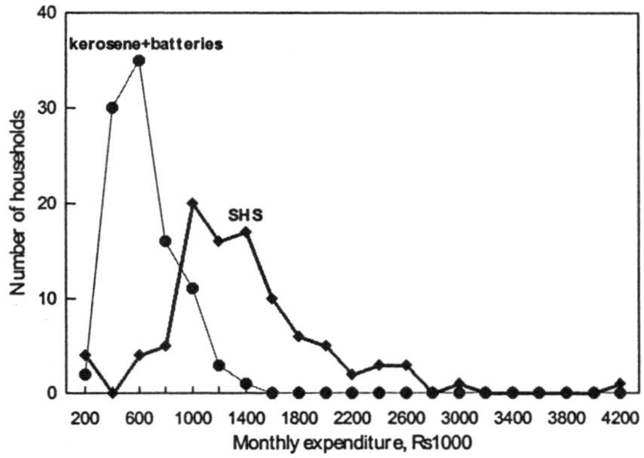

Figure 11.6 Monthly energy expenditures, before and after SHS

Figure 11.7 shows the relationship between disposable income and the size of the SHS: the relationship between the two is again weak (and not statistically significant). Although 11 households (that is, 11 per cent of the sample), reporting monthly incomes of less than Rs 3,000, purchased SHS, survey self-reporting of monthly disposable income is known to be generally understated.

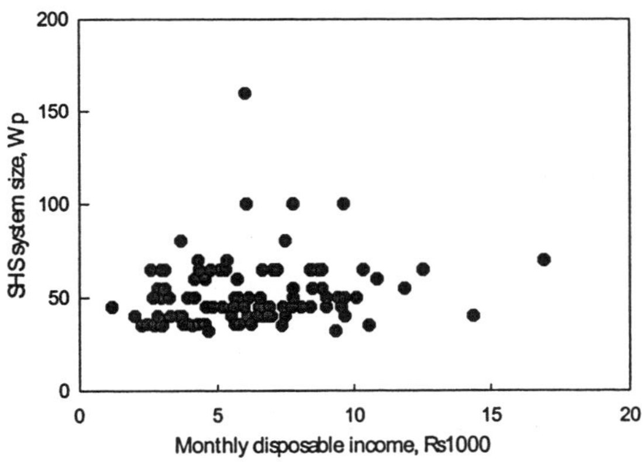

Figure 11.7 SHS system size v. disposable income

As expected, if one calculates the benefits only on the basis of avoided costs (of kerosene and dry cells), the economic returns of a SHS are negative. If we use the assumptions of the AC Nielson survey:

- average monthly expenditure on kerosene and battery charging prior to installation of SHS at Rs 495/month (Rs 5,940/year);
- four-year loan at 24 per cent interest, with 25 per cent down-payment
- discount rate 12 per cent;
- VAT on SHS at 11 per cent of retail price;
- tax component of kerosene price limited to National Security Levy (Rs 1.23/litre); and
- solar home system retail price Rs 46,000 (for a 45 Wp system).

then, as shown in Table 11.3, the economic rate of return is 7.7 per cent, significantly less than the probable opportunity cost of capital, while the financial rate of return to the homeowner is negative.

Table 11.3 Economic and financial returns based on avoided costs

	[unit]	NPV	1	2	3	4	5	6	7	8
costs of PV system										
Down payment	[Rs]	10268	11500							
GEF[to consumer]	[Rs]	0	0							
Govt.Grant[to consumer]	[Rs]	0	0							
Loan principal	[Rs]	30804	34500							
Finanancial cost	[Rs]	**41071**	**46000**							
Finance	[Rs]	38217		16905	14835	12765	10695			
Loan repayments	[Rs]	-38217		-16905	-14835	-12765	-10695			
LessVAT&duties	[Rs]	-4107	-4600							
Less income tax on margin	[Rs]	-5252	-5882							
Less transfers	[Rs]	-4877	-5462							
Economic capital cost	[Rs]	**26835**	**30055**	**0**	**0**	**0**	**0**	**0**		
O&M costs										
Bulbs	[Rs]	3539		582	582	582	582	582	582	582
Controller	[Rs]	2465						3104		
Battery	[Rs]	8566				4753			4753	
Financial cost to consumer	[Rs]	14570	0	582	582	5335	582	3686	5335	582
Less VAT	[Rs]	-1457	0	-58	-58	-534	-58	-369	-534	-58
Economic O&M costs	[Rs]	13113	0	524	524	4802	524	3317	4802	524
Total economic costs	[Rs]	**39948**	**30055**	**524**	**524**	**4802**	**524**	**3317**	**4802**	**524**
benefits at avoided costs										
Kerosene consumption	[litre:	1314		216	216	216	216	216	216	216
Kerosene	[Rs]	32118		5282	5282	5282	5282	5282	5282	5282
Battery&charging	[Rs]	4006		659	659	659	659	659	659	659
Dry cell	[Rs]	0		0	0	0	0	0	0	0
Hurricane lamp	[Rs]	0	0			0			0	
Petromax lamp	[Rs]	0	0				0			
Wick, gauzes	[Rs]	0		0	0	0	0	0	0	0
Total, financial	[Rs]	37437	0	5940	5940	5940	5940	5940	5940	5940
Kerosene duties	[Rs]	-1616		-266	-266	-266	-266	-266	-266	-266
VAT	[Rs]	0								
avoided costs	[Rs]	**35822**	**0**	**5675**	**5675**	**5675**	**5675**	**5675**	**5675**	**5675**
Net economic flows	[Rs]	**-4126**	**-30055**	**5151**	**5151**	**873**	**5151**	**2357**	**873**	**5151**
ERR	[ ]		7.7%							
net financial impact on consumers										
PV system	[Rs]	63054	11500	17487	15417	18100	11277	3686	5335	582
Replacement	[Rs]	36124	0	5940	5940	5940	5940	5940	5940	5940
Net flow	[Rs]	-26930	-11500	-11547	-9477	-12160	-5337	2254	605	5358
FRR	[ ]		-4.5%							

This result is consistent with those found elsewhere, when avoided costs are used as the measure of economic benefit: in the Philippines (Meier, 2003b), the ERR of a similar sized (40Wp) SHS was found to be 11.7 per cent, similarly below the hurdle rate (though in the Philippines, 15 per cent rather than 12 per cent is used).

Table 11.4 shows the corresponding reconciliation of economic and financial flows among the various stakeholders. Because the financial interest rate is 24 per cent, while the discount rate is 12 per cent, the difference appears as a residual benefit to the financial institutions. GEF is a source of funds, and therefore shows up as a financier; while Government is a net beneficiary, since the VAT levied on the SHS (paid by the dealers) significantly exceeds any VAT on kerosene lamps and battery charging, and the National Security Levy on Kerosene.

Table 11.4 Reconciliation of economic and financial flows (as NPV)

	Consumer	Dealers	Govt.	FIs	GEF	Total
Benefits	37,437		-1,616			35,822
Purchase price	-41,071	41,071				0
Loan	30,804					0
GEF Grant		6,790			-6,790	0
Govt. Grant						0
VAT (on equipment)		-4,107	4,107			0
Income tax (dealers)		-5,252	5,252			0
Income tax (FIs)			2,595	-2,595		0
Economic cost		-26,835				-23,835
Consumer finance	-38,217			38,217		0
O&M costs	-14,570					-13,113
Total	-25,617	11,667	11,795	4,819	-6,790	-4,216

The apparent negative financial return to homeowners is simply a reflection of the strongly negative cash flows in the early years as the loan is paid off (Figure 11.8).

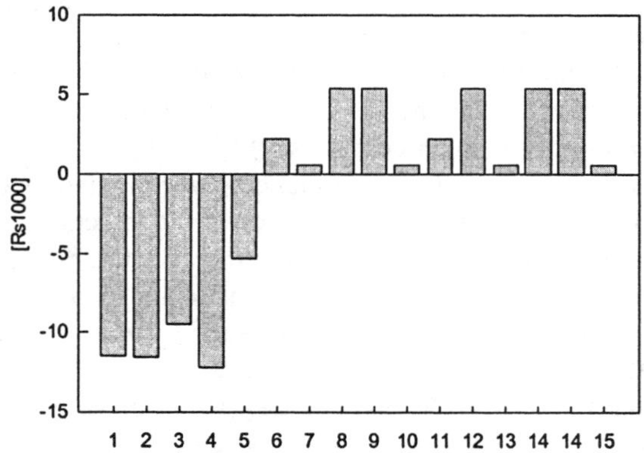

Figure 11.8 Cash flows to homeowners, Rs/year

It therefore necessarily follows that the economic returns must be higher than those assessed purely on the basis of the replacement costs – since consumers are in fact willing to pay the higher outlays for SHS. Using the approach outlined in the first section of the chapter, we may derive the demand curve for lighting as shown in Figure 11.9, as estimated for a homeowner purchasing a 45Wp system.

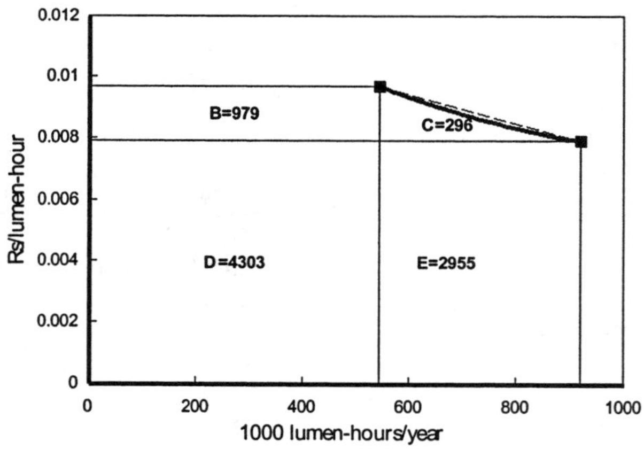

Figure 11.9 Demand curve for lighting (based on 45Wp PV system)

The corresponding set of consumer surplus calculations are shown in Table 11.5, with the distribution of costs and benefits as shown in Figure 11.10. The ERR is a robust 31%, which increases to 42.6% when the GEF contribution is added as an environmental benefit.

Table 11.5 ERR with consumer surplus benefits

		[unit	NPV	1	2	3	4	5	6	7	8
	Lighting costs, PV system										
[1]	Allocated costs		70%								
[2]	PV costs [financial]	[Rs]	44138	8050	12241	10792	12670	7894	2580	3735	407
[3]	..Levelised	[Rs]	44138	6329	6329	6329	6329	6329	6329	6329	6329
[4]	PV lighting	[lum	5593422	0	919800	919800	919800	919800	919800	919800	919800
[5]	P[PV]	[Rs/l	0.0079								
[6]	Area [D+E]	[Rs/y	7258								
[7]	***Lighting costs, Kerosene***										
[8]	Fuelcost	[Rs]		0	5282	5282	5282	5282	5282	5282	5282
[9]	Wick lamp	[Rs]		0	0	0	0	0	0	0	0
[10]	Petromax lamp	[Rs]		0	0	0	0	0	0	0	0
[11]	Wick, gauzes	[Rs]			0	0	0	0	0	0	0
[12]	Total costs	[Rs]	32118	0	5282	5282	5282	5282	5282	5282	5282
[13]	Kerolighting	[lum	3315928		545282	545282	545282	545282	545282	545282	545282
[14]	P[KERO]	[Rs/l	0.0097								
[15]	area [B+D]	[Rs/y	5282								
[16]	***Radio/TV: PV system***										
[17]	Allocated costs		30%								
[18]	PV costs [financial]	[Rs]	18916	3450	5246	4625	5430	3383	1106	1601	175
[19]	..Levelised	[Rs]	18916	2712	2712	2712	2712	2712	2712	2712	2712
[20]	PV non-lighting	[VL-	11098		1825	1825	1825	1825	1825	1825	1825
[21]	P[PV]	[Rs/	1.7								
[22]	Area [D+E]	[Rs/y	3111								
[23]	***Radio/TV: battery***										
[24]	total costs	[Rs]	4006	0	659	659	659	659	659	659	659
[25]	PV non-lighting	[VL-	1665		274	274	274	274	274	274	274
[26]	P[battery]	[Rs/	2.4								
[27]	Area [B+D]	[Rs/y	659								
[28]	**Net economic flows**										
[29]	***Lighting[see chart]***										
[30]	Total benefits [B+C+D+E]	[Rs/y	51887	0	8532	8532	8532	8532	8532	8532	8532
[31]	Total costs [D+E]	[Rs/y	44138	6329	6329	6329	6329	6329	6329	6329	6329
[32]	Net consumer benefits, lighting	[Rs/y	7749	-6329	2203	2203	2203	2203	2203	2203	2203
[33]	***TV/radio***										
[34]	Total benefits [B+C+D+E]	[Rs/y	26708		4392	4392	4392	4392	4392	4392	4392
[35]	Total costs [D+E]	[Rs/y	18916	3450	5246	4625	5430	3383	1106	1601	175
[36]	Net consumer benefits, TV/Radio	[Rs/y	7792	-3450	-854	-233	-1038	1009	3286	2792	4217
[37]	**Total consumer surplus**	[Rs/y	**15541**	**-9779**	**1349**	**1970**	**1165**	**3212**	**5490**	**4995**	**6421**
[38]		[ERR]		30.7%							
[39]	Total financial costs	[Rs/y	-63054	-11500	-17487	-15417	-18100	-11277	-3686	-5335	-582
[40]	Total CS benefit	[Rs/y	78595	0	12924	12924	12924	12924	12924	12924	12924
[41]	Net financial flows	[Rs/y	15541	-11500	-4563	-2493	-5176	1647	9238	7589	12342
[42]		[ERR]		21.1%							
[43]	Economic cost adjustment	[Rs/y	-0	-30055	4413	4413	4413	4413	4413	4413	4413
[44]		[Rs/y	15541	-41555	-150	1920	-763	6060	13651	12002	16755
[45]		[ERR]		16.8%							
[46]	..Less net govt. subsidies	[Rs/y	11795	13211							
[47]	..Less GEF subsidies	[Rs/y	-6790	-7605							
[48]	..Plus FI surplus	[Rs/y	4819	5397							
[49]	..Plus dealer surplus	[Rs/y	11667	13067							
[50]	Net economic flows	[Rs/y	37032	-17485	-150	1920	-763	6060	13651	12002	16755
[51]		[ERR]		31.0%							
[52]	With GEF as economic benefit	[Rs/y	43822	-9881	-150	1920	-763	6060	13651	12002	16755
[53]		[ERR]		42.6%							

Notes: Extract only, calculations done for the 15-year assumed life of the PV system.

The "economic cost adjustment" of rows [43]-[45] is required because without it, the series of economic flows has more than one turning point (because the subsidies and other adjustments occur in year 1), for which the ERR then becomes indeterminate. The NPV calculations are unaffected.

The adjustment for taking the GEF contribution as a benefit in year 1 is as per World Bank procedures. In effect one assumes that the global environmental benefit is exactly equal to this contribution, and hence benefits increase by this amount.

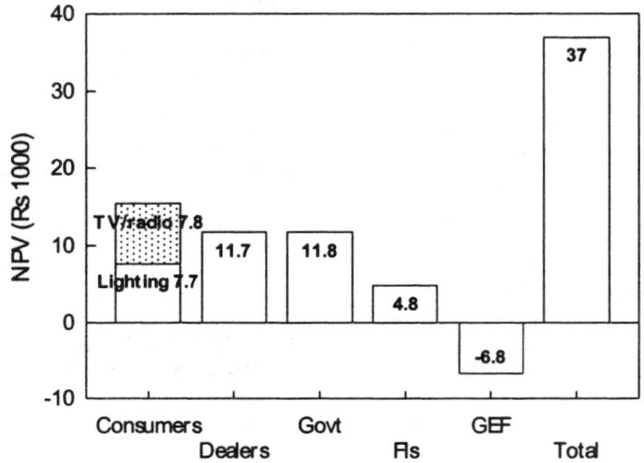

Figure 11.10 Distribution of costs and benefits

The implied willingness to pay for the tranche of electricity consumption provided by the PV system (for lighting and TV viewing) calculates to Rs 90/kWh (about US$1/kWh).[6]

In conclusion it seems reasonable to claim that SHS represent a viable approach to off-grid electrification, and that the subsidies provided by GEF are cost-effective in terms of levering economic benefits not otherwise achievable because of problems with financing.

Moreover, the demonstrated willingness-to-pay for SHS, in excess of replacement costs of kerosene, lends weight to the use of the consumer surplus methodology as a way of estimating benefits: use merely of kerosene replacement costs clearly underestimates the actual economic benefits.

11.3 VILLAGE HYDRO

Small micro-hydro systems were a second component of the ESD project, and again very successful: these are small hydro systems of less than 50 kW, implemented by village cooperatives. As part of the World Bank's evaluation of the ESD project (the so-called Implementation Completion Report, ICR), AC Nielson surveyed 100 village hydro system (VHS) beneficiaries, and 30 of the village hydro systems implemented under ESD. Unfortunately, generally inadequate record keeping by the village cooperatives makes precise determination of some of the key parameters difficult, such as achieved actual monthly and annual capacity factors.

At the time of appraisal, the ERR was estimated at (exactly) 12 per cent, which has the appearance of being contrived for the sake of meeting the 12% hurdle rate. However, evaluation of the systems actually implemented reveals lower costs, and therefore greater economic benefits than expected. For example, as against the estimated cost of US$ 2,023/kW,[7] the completed project costs reported by the Survey show an average capital cost of US$ 1,892/kW (Figure 11.11).

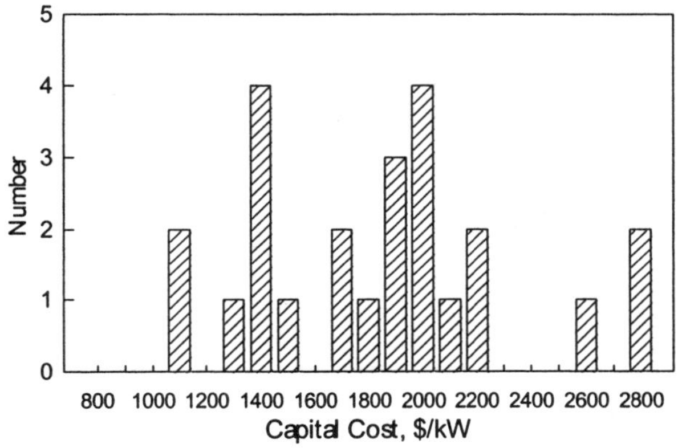

Figure 11.11 Village hydro capital costs

Figure 11.12 shows the relationship of unit capital cost against installed capacity, and confirms the strong extent of scale economies.

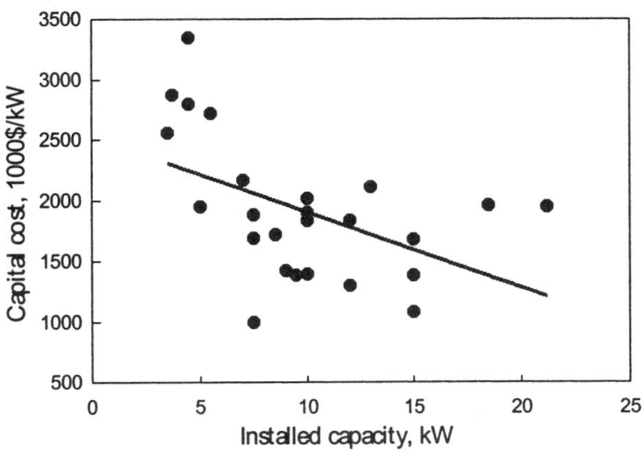

Figure 11.12 Capital cost v. installed capacity

In addition, there is a substantial 'sweat equity' component based on voluntary labour. Based on an average income of Rs 300/day, the survey report estimates the 'minimum opportunity cost of voluntary service rendered by the average VHS user at Rs 9,000 (30 man-days)' (AC Nielson, 2003). The beneficiary survey reveals that 20–30 days is the most frequent contribution of voluntary service.

Table 11.6 Estimates of sweat equity

System	Sweat equity	Cash equity	Total village equity	Project cost	Sweat equity (%)	Cash equity (%)	Adjusted cash equity (%)
	[1]	[2]	[3]	[4]	=[1]/[4]	=[2]/[4]	[2]/([4]-[1])
Handunella		125	125	1,325	0	9	9
Aluthgama	561	102	603	1,307	43	8	14
Diyapoda	280	134	414	1,152	24	12	15
Manamperigama	168	24	912	800	21	3	4
Dumbara	1,527	470	1,997	3,700	41	13	22
Average					26	9	13

Table 11.6 shows the estimated sweat equity component, based on the opportunity cost of contributed time (based on average household incomes). Taking the cash equity fraction at 15 per cent; making the conservative assumption that 25 per cent of the project cost is imported, then the economic capital costs (including sweat equity) average at US$ 2,060/kW, and the financial capital cost at US$ 2,225/kW (exclusive of sweat equity).

Records of kWh actually produced are sparse, notwithstanding instructions given to the operators by the developers and consultants that monthly readings of the kWh meter should be entered in the station logbooks. The sample survey did report estimated annual kWh, but the capacity factor estimates for many of the plants are 1.0, suggesting that these were not based on meter readings!

Indeed, actual readings were available for only 7 systems (column 3 of Table 11.7). The period of operation could be inferred from knowledge of commissioning dates to derive average annual kWh, and hence estimate the average capacity factor, with the results as shown in column 5. However, even in this case, the capacity factors estimated for two of the VHS are unlikely (at 0.23 and 0.14 much to low to satisfy consumers) and suggest that meters were reset to zero or replaced at some point since commissioning). Nevertheless, it seems reasonable to infer that the achieved capacity factors are in the range of 80–95 per cent.

Table 11.7 Capacity factor estimates

	Estimated annual kWh, survey	Estimated capacity factor survey	Actual meter reading (Spring'03)	Months	Estimated capacity factor (meter reading)
	[1]	[2]	[3]	[4]	[5]
Maddabaddara	35,040	0.889			
Kandaloya	74,256	0.848	352,716	56	0.863
Handunella	113,568	0.997	321,776	34	1.012
Gedarawat	76,752	0.730			
Golahinna	162,060	1.000			
Kaduoya	109,200	0.831	327,600	37	0.809
Kawudubukka	131,400	1.000			
Veediyawa	51,480	0.131	179,406	24	0.232
Samanala	148,512	0.997			
Ihalagonna	60,840	0.926	91,260	18	0.926
Glime	33,670	0.769	73,002	21	0.976
Hathkella	87,360	0.997			
Hinguralakande	65,700	1.000			
Gamperigama	4,269	0.139	2,490	7	0.139
Dumbara-Manana	24,768	0.133			
Wewagama	32,850	1.000			
Ritigala Ella	131,400	1.000			

These capacity factors are substantially higher than the 50 per cent estimated at appraisal. However, they are entirely plausible given the design philosophy, which is to match the size of the system to the dry season flows. For example, the feasibility study for the VHS at Handunella-Atulauda states[8]

> The average daily flow of the Ing Oya has been estimated at 241 litres/sec . . . and the continuous flow available throughout the year at 54 litres/.sec. The flow rate of 22.5 litres/sec will be taken as the design flow of the proposed hydro project. (Consultancy and Professional Services, 1999a).

Clearly, if a village hydro system were only operated at a 50 per cent capacity factor, then diesel backup would be needed to supply year-round power.

Although VHSs were built in small watersheds and generally well above gauging stations, none maintained records of gauged flows at the intake. Therefore, the likely monthly variation in flows can only be inferred from precipitation records that are generally available. Figure 11.13 shows the monthly variation in rainfall at the Ingoya Estate, on which the design flow for the Handunella-Atulauda VHS is based.

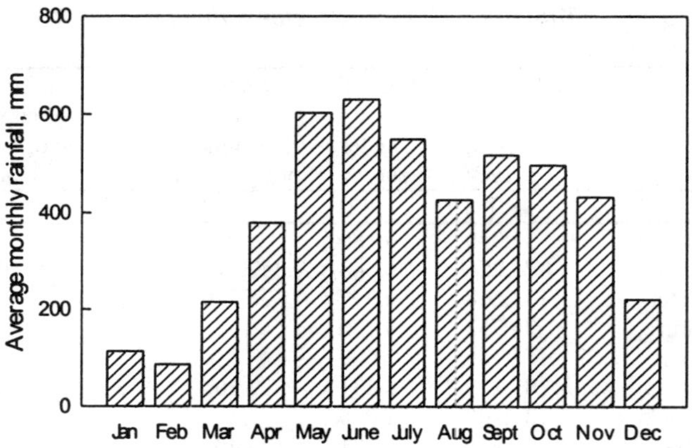

Figure 11.13 Monthly rainfall patterns, Ingoya Estate (mm)

If one matches the design flow to January/February conditions, high annual capacity factors are achievable. Nevertheless, there remains the question of the daily load curve: it seems unlikely that this continuously available power is actually consumed given the probable end-uses of lighting and TV.

Based on the VHS beneficiaries' average monthly expenditure on kerosene, batteries, dry cells of Rs 322/month; setting these equal to the benefit measure; and using the actually achieved capital cost of US$ 2,050/kW, then with the GEF grant taken as a benefit, the ERR for the same VHS (15 kW, 150 households) as used at appraisal works out at 21.8 per cent. Without the GEF grant component, the ERR is 17.9 per cent.

Table 11.8 Achieved ERR

		NPV	0	1	2	3	4	5	6
Capital cost	2050 ($/kW)								
KW	15.0 (kW)								
GEF grant	($US)	4,783		6,000					
Capital costs	($US)	-27,455	-30,750						
O&M costs	0.8% ($US)	-1,702		-6,255	-255	-255	-255	-255	-255
Total costs	($US)	-24,374	-30,750	5,745	-255	-255	-255	-255	-255
Capacity factor	0.800								
Benefit/kWh	5.7 (USc/kWh)								
Benefits	(MWh)			105	105	105	105	105	105
	($US)	39,960		5,992	5,992	5,992	5,992	5,992	5,992
Total flows	($US)	15,586	-30,750	11,737	5,737	5,737	5,737	5,737	5,737
ERR		21.8%							

Note that whether or not the high levels of electricity are actually consumed makes no difference to this calculation, since the benefits are simply taken as the pre-electrification energy expenditures: if smaller amounts of electricity were used, then the average benefit US 5.7 cents/kWh would simply be that much greater. Based on the survey results we may estimate the ERRs achieved for individual VHS systems as shown in Figure 11.14.

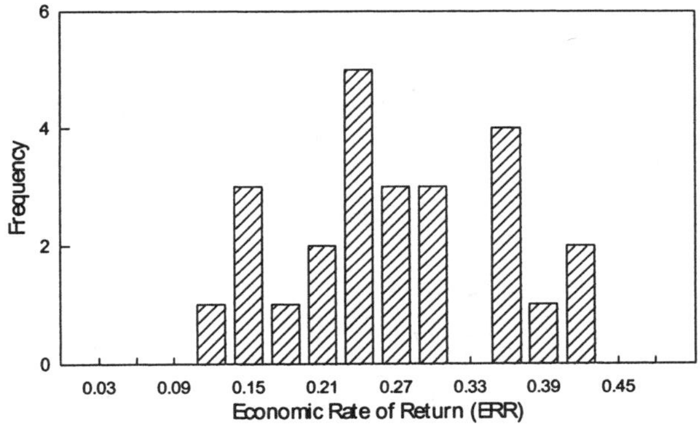

Figure 11.14 ERRs for surveyed VHS

These estimates of ERR based on replacement costs are likely to significantly underestimate the actual economic returns, because the willingness to pay (WTP) for the first tranche of electricity consumption is substantially greater than those implied by valuation of benefits as replacement costs. From the detailed analysis provided earlier for solar homes, we may infer that the first tranche of annual consumption of 115 kWh/year) may be valued at US$ 1/kWh.[9] If we assume further that the actual consumption is limited to evening TV viewing and lighting, and each household has 100watts of installed devices that run for 6 hours/day on average, then total annual consumption is $6 \times 100 \times 0.9 \times 365/1{,}000 =$ 197 kWh/year, implying a second tranche of 82 kWh/year. Then, even we if assume zero WTP for this second tranche, the ERR computes to 61.1 per cent, as shown by the calculations of Table 11.9 (or 53.9 per cent without the GEF grant).

Table 11.9 Benefits based on consumer-surplus

			NPV	0	1	2	3	4	5	6
Capital cost	2050	[$/kW]								
kW	15.0	[kW]								
GEF		[$US]	4783		6000					
Capital costs		[$US]	-27455	-30750						
Lost earnings[sweat equity]			-690	-773						
O&M costs	0.8%	[$US]	-1702		-255	-255	-255	-255	-255	-255
Total costs		[$US]	-25065	-31523	5745	-255	-255	-255	-255	-255
Capacity factor	0.800	[ ]								
Benefit/kWh	20.2	[USc/kWh]								
Benefits		[MWh]			30	30	30	30	30	30
		[$US]	115043		17250	17250	17250	17250	17250	17250
Total flows		[$US]	89978	-31523	22995	16995	16995	16995	16995	16995
		[ ]	61.1%							
Estimate of benefits based on WTP										
Consumption	6	[hours/day]								
Number of HH	150	[ ]								
	90	[watts/HH]								
		[wh/day/HH]	3601		540	540	540	540	540	540
		[kWh/year/HH]	1314		197	197	197	197	197	197
First tranche	1	[$/kWh]								
	115	[kWh]	767		115	115	115	115	115	115
		[$/year/HH]	767		115	115	115	115	115	115
Second tranche		[kWh]	548		82.1	82.1	82.1	82.1	82.1	82.1
	0	[$/kWh]								
		[$/year/HH]	0		0	0	0	0	0	0
Total WTP		[$/year/HH]	767		115	115	115	115	115	115
		[$/year]	115043		17250	17250	17250	17250	17250	17250

As always, such calculations abound with uncertainty, and economic rates of return should be viewed as probability distributions, rather than single deterministic numbers. In the Annex we explain how Monte Carlo simulations may be used to better estimate the likely probability distributions of ex-ante rates of return on projects.

Table 11.10 and Figure 11.15 show the distribution of costs and benefits. The bulk of the economic benefit accrues to the consumer, with some portion accruing to the village cooperative – though in practice, some of this surplus is likely to be passed back to the consumer beneficiaries (because of non-payment of dues, refunds in the event of cash surpluses, etc). The GEF grant is shown as a subsidy, assumed to be equal to the global environmental benefit. For government, the loss of National Security Levy (on kerosene) is roughly offset by taxes and duties on capital costs.

Table 11.10 Distribution of costs and benefits

	Consumers	Govt	VHS	GEF	Total	Global envt.	Total (incl. envt)
Benefits							
Electricity	115,043				115,043		115,043
Environmental					0	4,783	4,783
Fees	-39,974		39,974		0		0
Lost earnings	-690				-690		-690
Grants			0	-4,783	0		0
Capital costs			-27,455		-27,455		-27,455
O&M costs			-1,702		-1,702		-1,702
NSL, kerosene	2,216	-2,216			0		0
Taxes and duties on capital cost		2,344	-2,344		0		0
Total	76,594	128	8,472	0	85,195	0	89,978

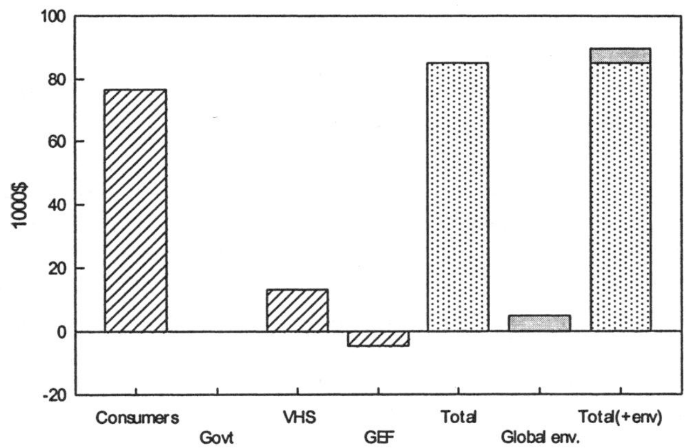

Figure 11.15 Distribution of costs and benefits

Box 11.1 Isolated mini-grids in the Philippines

Hydro is not always as economic as in Sri Lanka for isolated mini-grids. In the Philippines, diesel-based systems were found to be least-cost in all cases where a small hydro alternative was also identified. The table summarizes some of the key results of the feasibility studies for six communities on two islands for which grid extension was not judged economic.

	First year # of connected households	First year Installed capacity kW	ERR (diesel) (%)	Levelized costs US cents/kWh Diesel	Hydro
Roxas	1,441	625	**32.6**	15.0	26
El Nido	479	250	11.2	19.1	54
SanVicente	413	185	**19.7**	18.1	39
Taytay	1,216	575	**25.6**	16.1	
Palawan	3,549	1,625	**28.2**	15.3	
Jose Abas Santos	796	310	8.9	21.6	
Malita	1,778	685	**25.1**	16.3	23
Davao	2,575	990	**20.6**	17.6	41

Note: **Boldface** = economically viable (ERR>15 per cent hurdle rate)

Average WTP in the Philippines was estimated for a typical household consumption tranche of 300 kWh/year at US 30 cents/kWh. The figure shows the demand curve for the Philippines, whose points are from empirically established marginal WTP for different types of systems; the highest WTP is for the first tranche of about 30 kWh/month, as is typically provided by 20 Wp solar home systems; but thereafter, the WTP drops dramatically, falling to a typical 15 cent/kWh tariff charged in a Philippines mini-grid – which is a level confirmed by the observed willingness to pay for TV viewing of non-electrified households (by battery charging).

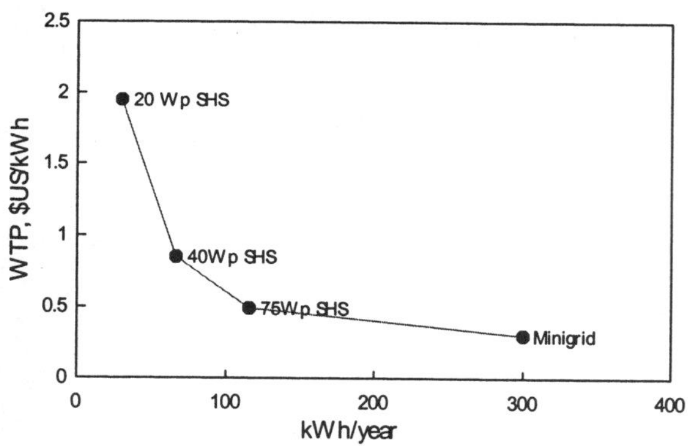

In summary, the achieved economic benefits are substantially greater than that estimated at appraisal. First, capacity factors are significantly greater in practice than as estimated at appraisal, a consequence of the design approach followed by the leading consulting firms and developers. This calls for systems to be sized on the basis of the available average dry season flow. Second, actual capital costs are slightly lower than estimated. Finally, expenditures for kerosene and battery prior to electric service have increased significantly since 1996 as a consequence of real income gains (and by more than the CPI), and consequently the willingness to pay for electric service is higher. From a purely economic point of view, then, village hydro systems bring significant benefits, and should continue to be supported as a viable option for off-grid electrification in areas where the grid access cannot be realistically achieved in the short to medium term.

ANNEX:

THE PROBABILITY DISTRIBUTION OF THE RATE OF RETURN

Simple calculations of an economic rate of return may suffice for a retrospective analysis to evaluate whether a project has met its original objectives. However, for a prospective analysis, that is, during project appraisal, when many assumptions are uncertain, and where there may be a number of project alternatives, more sophisticated approaches are required. One of the most common is so-called Monte Carlo simulation, now routinely applied in World Bank project appraisals for renewable (and other) energy projects. The idea is simple: each important input assumption is specified as a probability distribution (rather than a single "best" value). The ERR calculation is then repeated a large number of times (typically somewhere between 1,000 and 10,000), drawing at each calculation some sample value for each of the input assumptions. In this way a probability distribution of ERR is generated.

Figure 11.16 exemplifies the approach, taken from a recent analysis of small hydro projects in Vietnam to serve isolated communes not presently connected by the grid (Meier, 2003a). The alternatives are grid extension and powering by diesel – each of which is subject to its own set of uncertainties. Probability distributions for each of these alternatives are derived, as compared as shown.

This allows not just an assessment of the expected value of ERR – which in this case, for the Thong Thu commune, shows diesel to have a higher expected value – but also to examine the trade-off with risk, which we can express, for example, as the probability that the hurdle rate is not achieved (which is simply the area under the curve to the left of the 10 per cent hurdle rate).

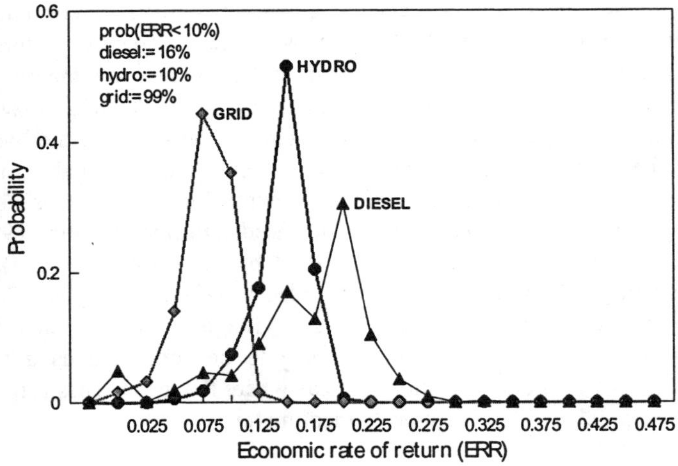

Figure 11.16 Probability distribution of the economic rate of return

The comparison of expected returns and risks for four communes is shown in Tables 11.11 and 11.12 clearly expected value alone does not tell the full story.

Table 11.11 Economic rate of return: ERR

		BatMoT	GiapTrung	ThongThu	PhuNam
Hydro	(%)	11.9	13.4	14.3	15.0
Grid extension	(%)	12.1	**14.7**	7.3	14.4
Diesel	(%)	**16.1**	12.8	**18.4**	**15.7**

Table 11.12 Risk of ERR<hurdle rate of 10 per cent

		BatMoT	GiapTrung	ThongThu	PhuNam
Hydro	(%)	27	**4**	**11**	**14**
Grid extension	(%)	20	8	98	20
Diesel	(%)	**18**	26	18	28

For example, in the case of Phu Nam and Giap Trung, all three alternatives have comparable expected value; but hydro has significantly lower risk, and so the hydro option would be indicated. In the case of Bat Mot, diesel has both the highest expected return (and lowest risk). In the case of Giap Trung, grid extension is indicated (since neither hydro nor diesel has a clear advantage), though the diesel option has the highest risk.

The point of this example is not only that a probability distribution of ERR is better than a single point (deterministic) estimate, but also that trade-offs between expected return and risk are often required.

NOTES

1 At the Burns Unit of the Lady Ridgeway Children's Hospital established in December 2000, by August 2001, 176 children had been treated for burn wounds. The majority of the victims were from the unelectrified rural villages of Chilaw, Puttalam, and Karapitiya that are dependent on kerosene for lighting (*Sri Lanka Sunday Observer*; September 23, 2001; Colombo).

2 A linear demand curve does not have a constant elasticity; in the case of Figure 11.1, for example, if the demand curve were a straight line between x and y, the own-price elasticity at point x is different to the own-price elasticity at point y.

3 Therefore, for the calculation of net benefits of electrification, there is no need to calculate the area A.

4 For example, a linear demand curve was used in the Philippines study of rural electrification benefits (Barnes et al., 2002). The World Bank's Rural Energy–I project in Vietnam also used a linear demand curve (ECA and Vernstrom, 2003).

5 Shell Solar Lanka has also conducted its own in-house surveys that report average monthly expenditures of 700–750 per month; however these may not be a random sample, but reflect households self-selected by the Shell staff who conducted the survey as the best prospects for a potential SHS sale.

6 A 75 Wp system generates about 315 watt-hours/day, or $315 \times 365/1{,}000 = 115$ kWh year, whose NPV over 15 years is 700 kWh. The NPV of the financial costs of the PV system is Rs 63,054, and hence the average WTP for the first 115 kWh per year is 63,054/700 = Rs 90/kWh.

7 At appraisal, the following assumptions were made

- Capital cost: US$ 2,023/kW
- O&M cost: 1 per cent of capital cost
- Discount rate: 12 per cent
- Installed capacity: 15 kW
- Capacity factor: 50 per cent
- GEF Grant: US$ 400/kW (US$ 6,000 for a 15 kW system)
- Number of customers per system: 150

The benefits were calculated at US 5.4 cents/kWh, based on a per household expenditure of Rs 104/month (at 1996 prices, and a kerosene cost of Rs 12/litre).

8 Similarly for the Pathavita project:

> The average daily flow of Kiruana Oya has been estimated at 636 litres/second and the continuous flow available for at least 11 months of the year had been estimated to be 100 litres/second. This flow of 100 litres/second will be taken as the design flow of the proposed hydro plant. (Consultancy and Professional Services, 1999b).

9 According to the AC Nielson survey, VHS beneficiaries have an average disposable income of Rs 6,963/year, slightly higher than that of SHS beneficiaries, whose income is reported as Rs 6,034/year. However, SHS beneficiaries were judged to be more likely to have under-reported income (10 per cent refused to reveal their income, as opposed to zero per cent of VHS beneficiaries). Therefore, it seems reasonable to believe that the WTP of VHS beneficiaries for the first tranche of high-valued electricity would be comparable to that of SHS beneficiaries. In any event, it is well established that there is a significant bias in self-reporting of income, that has been estimated by AC Nielson as 39 per cent; thus the adjusted disposable income figures increase to Rs 8,387 and Rs 9,678 for SHS and VHS beneficiaries, respectively.

12. Summary and implications for policy

This chapter presents a summary of the main results of our studies, and a review of key policy implications. In some cases (for example, unleaded gasoline) our recommendations have already been implemented by the Government of Sri Lanka, but in many other areas, the debate continues, some of which appear to be beyond the remit of the current Energy Sector Master Plan (as in the case of strategies to maximize carbon trading opportunities under the Kyoto Protocol). A new National Energy Strategy is now also under discussion, which takes up many of the themes discussed in this book. At the moment it is the power sector that is perhaps in greatest difficulty, but we doubt that restructuring and reform alone – though clearly a necessary condition – is sufficient to achieve the overriding objective of an adequate grid supply to consumers, at a reasonable price. We therefore close with an assessment of the main priorities.

12.1 THE POWER SECTOR

Sri Lanka will continue to pay the costs of the failures of decision-making over the past decade. Most of the generating capacity added over the past decade has been based on auto-diesel, which is not only the most expensive way of generating power, but also (as noted in Chapter 4) the most volatile in cost. Moreover, the environmental opposition to coal-generation has been largely irrational --: as we have shown in this book, the actual environmental impacts of diesel generation in densely populated urban areas is demonstrably greater than those of a modern coal-based plant. To be sure, the private sector has captured a significant share of Sri Lanka's power generation market, but the increased cost to consumers resulting from this use of diesel far outweighs the putative benefits of private sector competition in generation. In 2005 about half of all generation will be oil-based.

Fortunately, the discussions now underway to frame a new National Energy Strategy recognise this fundamental problem, i.e., Sri Lanka is now hooked on oil for power generation, and it is imperative from both a supply diversification and a cost perspective that the generation mix becomes more balanced. One proposal is to introduce a moratorium on oil-based plants, with a view to forcing in coal, and more renewable energy. Continued reliance on oil-based diesel generation forces upon consumers, very high electricity costs that will increasingly constrain Sri Lanka's international competitiveness. Most other countries, and certainly the successful economies of East Asia, have recognized this issue. Thus, moving towards a sensible generation mix must be one of Sri Lanka's most important energy-sector goals. Both the economic costs and local environmental benefits of replacing oil by coal exceed the benefits of generation privatisation.

GHG Mitigation

Undoubtedly the first priority for GHG emission reduction in Sri Lanka is power sector reform, which would undoubtedly result in a reduction of T&D losses, and the implementation of a comprehensive DSM lighting programme. Unlike reform in India, reform is win–win with respect to both net economic benefits and GHG emission reduction.

Once reform has been implemented, use of residual oil for diesel power generation represents the GHG mitigation option that would be easiest to implement from an institutional and managerial standpoint – and indeed will likely bring economic benefits as well. However, the availability of this option (and its economic benefits) depends upon an expansion of Sri Lanka's domestic refinery capacity, the feasibility of which is still under study.

The least-cost option is determined to include additional DSM and T&D loss reduction (beyond levels as might be envisaged in a first step of reform), plus an aggressive implementation of some 45 MW of mini-hydro schemes. However, although these are 'win–win' in terms of economic and GHG emission reduction benefits, the financial and institutional barriers are significant. For this reason efforts to implement mini-hydro schemes should continue to be supported by GEF and the ESD.

Dendro-thermal power plants offer substantial potential for GHG emissions reductions at a reasonable incremental cost of avoided carbon. However, many of the assumptions can only be verified by field demonstration on a sufficiently representative scale to be able to answer critical questions about plantation management, land and labour availability, and plant-gate cost.

It is therefore recommended that a demonstration plantation be established at a scale to provide fuelwood for at least a 1 MW unit – which requires about 400 ha. In its initial stages, the output of such a project could be marketed as fuelwood in Colombo.

If a decision is made to proceed with the coal-fired unit on the

West Coast, we recommend that a study be made of the incremental costs of providing the second 300 MW with co-firing of fuelwood. This involves tested technology, and is already practised in many places worldwide. The timing of any such commitment to co-firing could be based upon a successful demonstration of the plantation concept.

Undoubtedly the more important constraint to the dendro-thermal option is the capital cost of the gasifier. The experience of the GEF demonstration project in Brazil suggests that gasifier costs may be substantially greater than US$ 1,200/kW, and reach costs in the range of US$ 1,200–1,500 only after 5–10 years of large-scale implementation. If this were indeed the case, then the best strategy for Sri Lanka would be to await the benefits of this experience in larger countries, and use co-firing at the coal plant in the interim.

In any event, even if co-firing were to prove too expensive (or otherwise undesirable from the point of view of any IPP operator of the coal plant), the wood from the demonstration plantation could still be sold in the Colombo fuelwood market. Therefore the risks of such a demonstration programme would appear manageable.

With respect to other renewable energy options in the power sector, Sri Lanka has substantial wind power potential, but the costs are high: the maximum wind scenario examined in this study has a cost of avoided carbon of US$ 65/tonne. Even if capital costs were to fall to US$ 700–800/kW (from the present US$ 1,175/kW), as forecast by some wind power advocates, this would still be an expensive mitigation option -- given the low plant factors achievable. The disappointing wind speeds and capacity factors achieved to date at the 3 MW demonstration project on the South Coast underscore the need for a more comprehensive wind resource assessment, before committing to large-scale wind development.

There are many reasons for implementing the solar homes programme given the importance that should be attached to rural lighting (and the expense and difficulties of extending grid service to remote villages). But as a GHG mitigation measure it would have low priority. Even under the highly optimistic assumptions about market penetration, the quantity of GHG emissions reduced by the SHS programme is less than 1 per cent of the reform scenario emissions.

Hedging strategies

High sulphur fuel oil can be imported at relatively low cost from Middle Eastern refineries, and over the past decade there have been many periods when this was a cheaper fuel than coal. However, given the present volatility of oil prices (which shows no sign of moderation), and the experience of the 1970s oil shocks, whose effects lasted until January 1986, the Government of Sri Lanka might be quite cautious about any commitment to a large steam-cycle oil plant. Indeed, while the historical dependence on hydro insulated the electric sector from the 1970s oil price shocks, the risks of over-dependence on hydro were amply demonstrated during the 1996 drought which caused severe dislocations to the economy. Thermal generation reduces the potential impact of this hydro risk, but one may hardly be eager to replace it with a new dependence on oil.

However, one useful hedging strategy is dual-firing capability at the coal plant – which may be provided with oil or wood co-firing. Assessment of the incremental capital costs of boiler modification may be straightforward. However, more complicated is the question of the amount of oil storage to be provided. The larger the volume of storage, the greater the ability to take opportunistic advantage of periods of low oil prices, and the less is the need for futures-market hedging. On the other hand, the larger the volume of storage, the greater the up-front capital cost. We urge that this be examined in detail for the forthcoming coal project.

12.2 PRIORITIES FOR THE TRANSPORTATION SECTOR

The highest priority is rationalization of fuel and vehicle taxation. If fuel taxes are to be equalized on the basis of revenue neutrality, then from an environmental perspective there is a slight preference for an equalization based on carbon emissions (which would give diesel a slightly higher rate of tax than gasoline on a Rs/litre basis). However, this is a relatively minor issue compared to the more important one of reducing the gap between diesel and gasoline taxes. Indeed, based on road damage externalities, there is an economic case for the tax on diesel to be greater than that on gasoline.

There is little evidence that the present tax structure is progressive, and that the high tax on gasoline serves the social objective of income redistribution. In fact, a significant proportion of gasoline taxes fall on the 700,000 motorcycle and three-wheeler owners. Our analysis shows that it is unlikely that private owners of gasoline-powered cars account for more than 30 per cent of vehicle-km (in gasoline cars), since the overwhelming majority of vehicles driven at high utilization are taxis, tourist vehicles, or are owned by businesses. The new petroleum pricing system introduced in 2002, with its better alignment to international oil price levels, is a great improvement over the previous *ad hoc* system. However, the relative prices of diesel and gasoline remain unsatisfactory, and do not reflect actual road damage and environmental externalities.

Reducing Sulphur in Auto-diesel

As in the case of leaded gasoline, the actual sulphur content in diesel has been significantly lower than the 1.1 per cent of the specification. Nevertheless, the benefits of reducing sulphur in diesel fuel are still significant. On the cost side, the best indication is the differential between gasoil of various sulphur content as revealed by the Singapore spot markets: the difference between 1 per cent S and 0.5 per cent sulphur gasoil has generally been in the range of 10–20 US cents/bbl. However, these differentials are trivial compared to the variation in overall price levels (that track the world oil price).

Given that the cost of modifying the kerosene unifiner for hydrotreating of gasoil are now sunk (with completion expected shortly), the refinery is now in a position to produce significant quantities of 0.1 per cent S gasoil, which can be blended with imports to almost any desired specification.

There is little doubt that a 0.5 per cent sulphur specification is cost-effective from a social standpoint, using highly conservative values for the health damage of SO_2. If the damage values suggested by the recent World Bank 'Six Cities' study were used, reduction to 0.3 per cent sulphur would also be cost-effective.

LPG Conversion of Gasoline Cars

Although the rate of growth of LPG conversion of petrol vehicles has slowed over the past year, we estimate that presently some 10 per cent of the gasoline-engine fleet runs on LPG, of which over 70 per cent involve lower-cost conversions to use bottled gas. This conversion is driven entirely by the distortions of fuel taxation policy: the bulk of the conversions appear to be high-mileage vehicles, notably taxis, for which conversion costs have (at least in the past) been highly profitable (in financial terms).

We find that the main environmental benefit of LPG conversion is the avoidance of lead emissions from leaded gasoline – a benefit that is now moot with the introduction of unleaded gasoline. Any remaining environmental benefits are dwarfed by the high economic costs of vehicle conversion and the higher economic costs of imported LPG and filling station costs. This is true even if the new 'tanktainer' concept were implemented.

There is significant loss of tax revenue to the government from LPG conversion (given that gasoline is highly taxed, whereas LPG incurs only the 6.5 per cent National Security Levy). This is effectively a subsidy to taxi fleets (and their high income ridership) and to businesses for whom the conversion costs are justified by high mileage.

The obvious remedy would be to tax auto-LPG at filling stations at a rate commensurate with gasoline, but this is unlikely to be effective since it would simply encourage use of bottled gas (unless the tax on bottled gas was also increased, which may be politically difficult). And efforts to prohibit bottled gas for auto use would surely fail, given the large number of vehicles that would be affected.

In fact, the poor overwhelmingly still use fuelwood for cooking, and bottled gas is largely used by middle and high income groups. These are the same groups whose predominant transportation modes are motorcycles and cars. Therefore, if a revenue neutral tax reform were undertaken (by equalizing tax on LPG and gasoline), the distributional impacts would cancel out, with the main benefit that Sri Lanka avoids the uneconomic use of LPG as a transport fuel. However, these conclusions may not necessarily extend to LPG conversion of diesel vehicles, notably buses – which has not heretofore been done in Sri Lanka.

Transportation Infrastructure Projects

Two of the transportation infrastructure projects (the airport expressway, port inland clearance depot) are win–win, while railway electrification, as noted above, is uneconomic, and cannot be justified by the (small) environmental benefits. However, as shown in Table 12.1, the proportions of environmental benefits of the Expressway and the dry container port in the total benefit are quite small (3 per cent and <1 per cent, respectively): inclusion of the environmental benefits simply make already good projects slightly better.

With respect to our negative conclusions on railway electrification, this applies only to the specific scheme examined, and may not apply to other proposed schemes. In particular the option examined here differs from so-called "light rail" systems, which are less capital intensive than mainline electrification. The light rail option will be examined in the next phase of research. The congestion mitigation options (reintroduction of school buses, congestion fee) are also win–win, though again the environmental benefits are modest.

Table 12.1 Percentage of environmental benefit in total social benefit

Option	Percentage
Congestion Fee	1
School Bus	5
Tax Equalization	30
Unleaded gasoline	120
Carbon tax (incremental)	40
Carbon tax (equalization)	31
Ban 2S MC	135
CKE (airport expressway)	3
Port inland clearance depot	0
Railway electrification	2
Sulphur 0.5 per cent	116
Sulphur 0.25 per cent	130
LPG (v. leaded gasoline)	472
LPG (tanktainer)	148
LPG (v. unleaded gasoline)	0

Note: Where the percentage figure exceeds 100%, it means that the option has a negative NPV before environmental considerations are taken into account. For example, in the case of a ban of 2S motorcycles, the value of the (positive) environmental benefits are 135% of the (negative) non-environmental benefits (i.e., of the non-environmental costs).

Table 12.2 displays these results in a different way. The local environmental benefit multiplier shows by how much the damage valuations could decrease and still justify the project on environmental grounds. For example, in the case of unleaded gasoline, the damage valuation could decrease by a factor of 5.9 before net benefits were zero. Thus, the higher its value, the more robust the conclusion that the environmental benefits justify the policy. On the other hand, where this multiplier is shown as a positive number, then it shows by how much environmental damages would need to increase for the project to have a positive NPV – in the case of railway electrification, environmental damage values would need to increase by a factor of 20.3 – which is sufficiently high for us to reject this option.

Table 12.2 Economic v. environmental benefits, NPV (discount rate 10 per cent), (Rs Million)

	NPV	Local	Multiplier	Global	NPV
Congestion Fee	7,616	50	win–win	21	7,687
School Bus	1,247	61	win–win	33	1,341
Tax Equalization	8,920	4,820	win–win	2,073	15,813
Unleaded gasoline	-3,297	19,549	-5.9	34	16,285
Carbon tax (incremental)	1,340	1,279	win–win	601	3,220
Carbon tax (equalization)	9,100	4,989	win–win	2,143	16,232
Ban 2S MC	-1,436	4,463	-3.1	269	3,296
CKE	12,281	333	win–win	574	13,188
Inland container port	8,642	31	win–win	12	8,685
Railway electrification	-9,478	182	20.3	27	-9,218
Sulphur 0.5 per cent	-467	3,360	-7.2		2,892
Sulphur 0.25 per cent	-1,290	5,600	-4.3		4,309
LPG (v. leaded gasoline)	-3,213	4,077	-1.3		864
LPG (tanktainer)	-1,315	4,077	-3.1		2,762
LPG (v. unleaded gasoline)	-3,213	0			-3,213

The most attractive options – that have the highest NPVs – are tax equalization (whether by volume or by carbon content), the introduction of unleaded gasoline, and the Expressway (CKE). However, in terms of environmental benefit achieved, by far the most attractive is the introduction of unleaded gasoline. With a multiplier of 5.9 (based on a health benefit of Rs 7/litre of gasoline), this would still be attractive even if the health benefits were reduced to just Rs 1.2/litre – which is two orders of magnitude lower than the value suggested by the benefit transfer method. One may therefore be reasonably confident that the option should be undertaken. Moreover, this does not include the additional benefits derived from catalytic converters (that are presently disconnected from reconditioned Japanese vehicles).

Figure 12.1 displays these results in the form of a trade-off plot: the line shows the options in the non-inferior set (that is, those that define the best set of options taking both objectives into account).

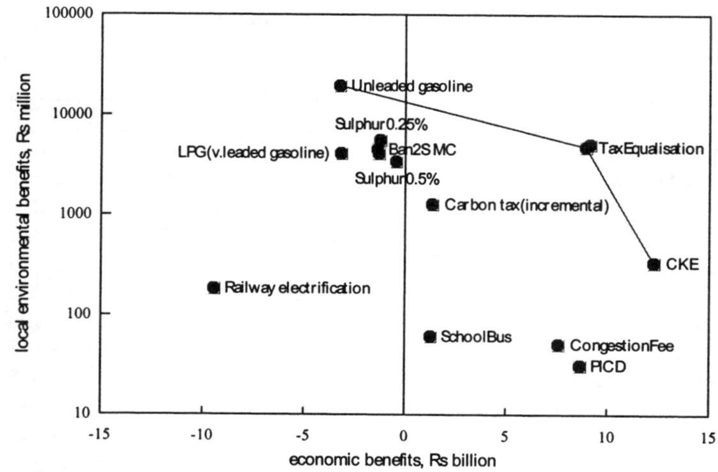

Note: Log scale for Y axis

Figure 12.1 Trade-offs, economic v. local environmental benefits

Figure 12.2 shows the corresponding figure for global benefits, based on a carbon price of US$ 20/tonne. Unleaded gasoline falls from the trade-off set, as expected: again the CKE and tax equalization are the best options.

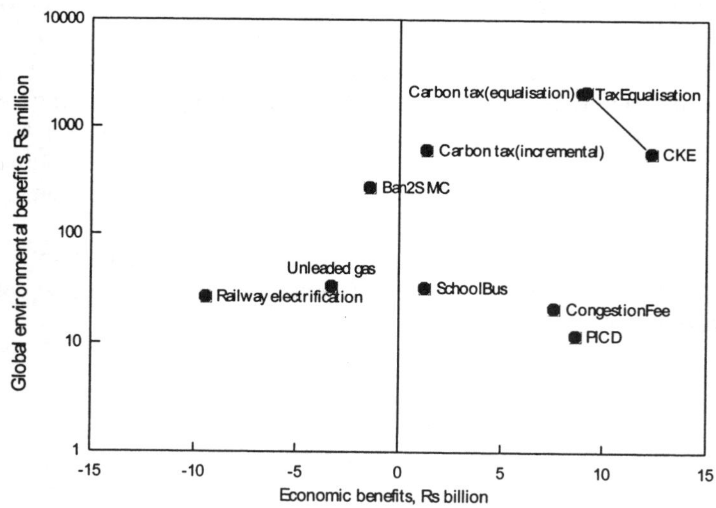

Figure 12.2 Trade-offs, economic v. global benefits (monetized at US$20/tonne C)

Finally, Figure 12.3 shows the trade-offs between economic benefits and PM–10. Banning the importation of two-stroke motorcycles is seen to be the most effective option for reducing particulates, though at some economic cost.

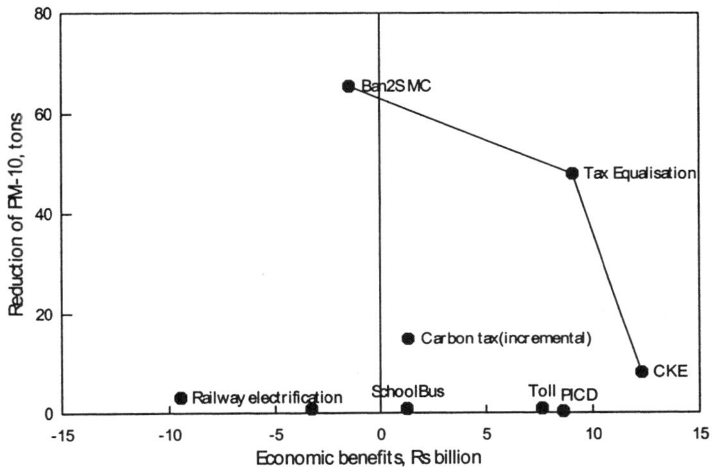

Figure 12.3 Trade-offs: economic benefits v. PM-10

12.3 LOCAL ENVIRONMENTAL IMPACTS

Given the number of recent detailed health damage studies in cities of other developing countries, it is clear that health damage valuations used in earlier studies in Sri Lanka are of an order of magnitude too low as regards vehicle emissions (or those arising from self-generated electricity). While the benefit transfer method used in this book provides reasonable indicative estimates of health damage costs, there is no substitute for local studies. The quality and consistency of monitoring data at the two NBRO sites since 1997 is good, but needs to be expanded to at least a third far-rural site to provide a better indication of background levels. A rigorous damage cost study using Colombo medical and hospital data remains a high priority.

There is evidence that air quality in Colombo has deteriorated over the last few years, and particularly in the winter of 2000/2001 the number of exceedances (of both daily average or daily maxima) has increased sharply. However, the most likely explanation of this deterioration is the start-up of the 60 MW barge-mounted power plant located in Colombo harbour, arguably the worst possible location for a power plant burning heavy oil. While there is evidence that traffic congestion during peak hours has also increased sharply over the past four years, the resulting increase in fuel consumption (consequent to lower fuel efficiency at lower speeds) does not explain this worsening of air quality.

Sulphur dioxide emissions are directly proportional to the quantity of diesel fuel used: and the fuel oil used in the barge-mounted plant is roughly 2.5 per cent sulphur by weight, or over three times the present sulphur content of auto-diesel. Indeed, the quantity of sulphur released by the barge-mounted plant is 16.5 tonnes/day, as opposed to 21.5 tonnes/day from auto-diesel used in transportation in the Colombo area. Even if the barge-mounted plant were dispatched only 8 hours/day, its operation would still represent an increase in sulphur emissions of 25 per cent – upwind (for a majority of hours of the year due to strong sea breezes) of the most densely populated area of Sri Lanka.

A comparable calculation for the proposed coal-based plant at Puttalam is equally instructive. If the coal-plant had been constructed (as recommended by every CEB generation plan since the late 1980s), it would be reasonable to suppose that the first 150 MW unit would now be in operation. Its daily emissions of SO_2, if using low-sulphur (0.6 per cent by weight) imported coal, without flue gas desulphurisation, compute to 7.86 tonnes per day, that is, 50 per cent of the emissions of the 60 MW barge mounted plant! Moreover, recent studies by the World Bank show that the typical ratio of health damages per tonne of emission from an urban, low stack source, compared to those of a remote, high-stack source are between 5 and 10 times greater. Thus per kWh generated, the probable health damages from the barge-mounted plant are 20–40 times greater.

There is also no case for flue gas desulphurisation (FGD) at Puttalam. The estimated damage cost per tonne of SO_2 per million population per day is calculated at US$5560 for the barge-mounted plant, US$46 for (300 MW) coal at Puttalam, and US$10,470 for Colombo's auto-diesel emissions. The reduction of 90 per cent of Puttalam's coal emissions by FGD would reduce health damages in Colombo by US$41/day! We may therefore conservatively conclude that:

- The health damages from the 60 MW barge-mounted plant are an order of magnitude greater than are the emissions from a coal-based plant at Puttalam;
- The emissions from the barge-mounted plant add 25–50 per cent to the auto-diesel emissions in the entire western province (depending on how the plant is dispatched);

- The environmental objections to the coal plant at Puttalam are completely irrational given that the probable health damages of oil-based plants are of at least an order of magnitude greater;
- The argument that the coal plant requires an FGD system is even more irrational. Reducing sulphur in diesel oil from the present 0.7 per cent to 0.5 per cent is 100–1,000 times more cost-effective than installing FGD at Puttalam.

All of the measures examined have net environmental benefits (when emissions are monetized), though some options involve trade-offs between emissions. Thus a ban on further imports of 2S motorcycles would (significantly) reduce particulate emissions, though increase NOx emissions: but since PM–10 has far higher damage costs, there is still a net reduction in health damages.

As shown by the example of railway electrification, the traditional focus on emissions is misplaced. Given that the electricity for electrification would come from coal-based power generation, sulphur emissions would increase (even if low sulphur coal, 0.6 per cent by weight, is used); but the resulting health damages would decrease, because the location of the sulphur emissions changes from densely-populated areas at more or less ground level (from DMUs) to tall stacks at remote locations (for the coal plant). By the same token, if the incremental electricity were generated from heavy fuel oil at a Colombo urban location (for example, at the location of the barge-mounted plant, or at Kelanintissa or Sapugaskanda), not only would sulphur emissions increase even more, but the damage costs would be equivalent.

12.4 THE PETROLEUM SECTOR

The introduction of competition in the petroleum sector is to be welcomed. However, as in the case of the power sector, the presumption that all can now be left in the hands of the private sector and to the interplay of market forces, carries grave long-term risks. The history of international oil markets over the past 50 years illustrates the need to balance short-term market considerations with the security of supply.

The future of the Sapugaskanda refinery is a case in point. There is no requirement in the present agreement between the Government and LIOC that LIOC should purchase some or all of its supply from the local refinery. If the restructuring of the sector proceeds as planned with the entry of a third player, the market share of Ceypetco (CPC) may drop below 30%, with the result that unless the other players purchase a significant share of their requirement from Sapugaskanda, the refinery throughput will fall to levels at which it is no longer profitable.

Clearly, it is not in the interests of Sri Lanka to continue running an unprofitable refinery, nor is it economically efficient for the refinery to be protected by customs duties (which will simply raise the price to *all* consumers). However, were the refinery to close, then Sri Lanka becomes entirely beholden to the international players that dominate the Asian marketplace – which is precisely the situation faced by Sri Lanka in the 1950s, and which contributed to the rationale for nationalisation at that time.

Several options to address this problem may be considered. For example, if the Sapugaskanda refinery can match the landed cost of products as calculated by the new pricing formula (based on Platts), then the private sector players could be required to purchase locally. To the rejoinder that this might deter potential new private sector entrants, or be seen as onerous by LIOC, it should be noted that privatisation should not simply be an end unto itself, but needs to be seen in the context of benefits to the national interest. In short, these kinds of issues can only be considered in a national energy planning process that balances the various long and short-term interests, including the security of supply. All stakeholders, including the private sector companies, should be represented in the process, but the final decision should remain the institutional responsibility of the Ministry of Power and Energy.

12.5 RENEWABLE ENERGY FOR OFF-GRID ELECTRIFICATION

There can be little doubt that the Energy Services Delivery (ESD) project is one of the most successful renewable energy programmes in the World Bank/GEF portfolio. In addition to the SHS and village hydro components evaluated in Chapter 11, ESD included a grid-connected mini-hydro programme, which also achieved its main goal, namely the establishment of a viable and sustainable private sector business activity developing small hydro projects.

We argue that consumer surplus be used as a measure of the economic benefits for off-grid electrification systems. Providing one is mindful of the pitfalls (e.g., avoid linear demand curves that result in over-estimation of benefits), the methodology is supported by the empirical evidence: even poorer households are willing to pay substantially more for the higher service quality of electric-based lighting and TV viewing than the kerosene/battery replacement cost.

To be sure, grid-based electrification remains the long-term objective, but as shown throughout South Asia, unrealistic rural electrification targets often result in great financial difficulties for the distribution companies, and grids are extended more for political objectives than for sound economic reasons, much less with a well structured subsidy policy in place. The inevitable result is high losses, poor grid service for everyone, and unsustainable financial losses for the utility. Throughout Asia, new models for rural electrification are being developed, the common theme of which is the transparency of subsidy and clear separation of O&M operations. Whether rural electrification is grid or off-grid based, O&M in distant rural areas requires effective local institutions, not a branch office of a remote organisation headquartered in a capital city.

12.6 THE MAIN PRIORITES

Energy planning, particularly when conducted by the public sector, is unfashionable and much maligned at a time when the received wisdom is market liberalisation and private sector involvement. Yet Sri Lanka's experience suggests that energy planning can be an effective instrument for guiding change and mobilising capital, provided it is coupled with the necessary political will.

The achievements of the institutions established by the national energy planning effort of the early 1980s, at a time when President Jayewardene also held the position of Minister of Energy, speak for themselves. Notwithstanding the predictions that public ownership of LECO would doom its efforts to meet its mandate for providing commercial service to its customers, LECO is a model for all of South Asia. It is true that LECO is not burdened with rural electrification operations, and has a relatively compact service area, but bringing down T&D losses from over 50 per cent to less than 8 per cent was a remarkable achievement under the difficult circumstances faced by the LECO management (including years of civil strife, a poor tariff environment, and disputes with CEB over large industrial customers).

However, the key to LECO's success was not merely a mandate for commercialisation of operations, but the political will – of a President elected by a substantial majority – to enforce that mandate. In practice, this meant direct support from the President when political pressure was put on the LECO General Manager to relax enforcement of those measures introduced to reduce commercial losses.

The experience of the planning cells of the main sector institutions established in the mid-1980s may be viewed as less successful, especially in the case of CEB. However, it is quite clear that the failure to ensure adequate capacity since the mid-1990s, and the increasing dependence of the CEB system on the most expensive form of power generation (based on auto-diesel) is not a failure of planning, much less a failure of nature to provide good hydro years, but a simple failure of political will to implement plans in a timely way.

Sri Lanka has been ill served by advice from those with a vested interest in privatisation, and particularly from those with vested interest in private sector generating plants. As in India, the proposition that the introduction of IPPs would lead the way to reform was always nonsense (and, at least in the case of India, promoted in the early 1990s mainly by those seeking opportunities for corruption). While the scale of failures in Sri Lanka is thankfully much smaller than at Dabhol, the result is much the same – continued power shortages, power generation plants using the most expensive possible fuel (auto-diesel), located in the worst possible places from an environmental point of view (such as the barge-mounted plant in Colombo harbour), and wishful thinking at its worst as vested interests (including bilateral assistance agencies for whom privatisation at any price was the operative mantra) promoted utterly unrealistic and unbankable projects (as in the case of the Mahaly coal-fired project at Trinicomalee).

Yet the presumption that the problems of the power sector will be solved by restructuring is also naïve, and the proposition that competition and the marketplace alone will ensure timely investments of new capacity is deeply flawed. While the California power crisis of 2000–2001 is perhaps an extreme case of failed deregulation (see World Bank, 2001, for a discussion of the lessons for developing countries), experience elsewhere shows the dangers of the naïve belief in the marketplace (as in the case of the inadequacy of the power transmission infrastructure in Europe, that in September 2003 caused widespread and costly blackouts in Italy). Because of very long gestation times in the power sector, there simply is no alternative but for the relevant regulatory body to insist on a coherent planning process by which to gauge and guide the adequacy of investment plans in the public interest. This imperative has been recognised in the discussions now underway to frame a new National Energy Strategy, and it has been proposed that a regular long-term generation expansion plan for the electricity sector should be prepared, and updated not less than once in two years. This process is to be complemented with a parallel renewable energy plan to be facilitated by the Energy Conservation Fund.

The Indian experience with the newly established electricity regulatory commissions shows that, notwithstanding their nominal independence and the unbundling of sector operations, political interference remains alive and well. Fortunately Sri Lanka is not afflicted with some of the more egregious political compulsions to the north, notably 'free power' for farmers, but one may confidently predict that in Sri Lanka the actual pace of change on the ground will be slow.

While sector reform, and a predictable and stable tariff regime, are necessary conditions for successful mobilisation of private capital for major energy infrastructure projects, the real constraint is the uncertainty of the 'peace' process. Alas, as with so many aspects of Sri Lanka's future economic development, the overriding priority for the energy sector is a sustainable settlement of the ethnic conflict. All else is secondary, and to assume otherwise is wishful thinking.

Bibliography

Abeygunawardena, P., *et al.*, (1999), 'Yunnan Dachaoshan Power Transmission Project', in *Environment and Economics in Project Preparation: Ten Asian Cases*, Manila: Asian Development Bank.

Abeywardena, A. (1988), 'National Energy Planning in Sri Lanka', SLEMA Journal, **2** (1), 3-9.

Abeywardena, A. (2002), 'Transport fuel pricing and taxation in Sri Lanka', presented at the SARI Seminar on *Trade in Cleaner Fuels,* Colombo, March 9-12.

AC Nielson (2003), *Statistical Baseline Survey of Renewable Energy Projects*, Colombo.

ASCI (Administrative Staff College of India) (1999), 'India: Global Overlay for Andhra Pradesh', Report to the World Bank, New Delhi.

AEA (1998), 'Generation and Environment, UK Perspective', Report AEAT 3776.

ASEAN (2001), 'Proceedings of the Regional Workshop on Fuel Quality and Alternative Fuels', Delhi, May.

Awegoda, C.M., M.S. Perera and D.T. Mathhes, (1994), 'An assessment of blood lead levels of the population exposed to vehicle emissions', Sri Lanka Academy of Science, Abstracts, E2.

Awerbuch, S. (2000), 'Getting it right: the real cost impacts of a renewables portfolio standard', *Public Utilities Fortnightly*, 15 February.

Awerbuch, S., and M. Berger (2002), 'Energy diversification and security in the EU: mean-variance portfolio analysis of electricity generating mixes and its implications for renewables'. IEA Technical Report, Paris.

Balce, G. (2001), 'Overview of regional fuel market and fuel quality standards' in Proceedings, ASEAN Regional Workshop on Fuel Quality and Alternative Fuels, Delhi, May.

Baldwin, M. (Ed.) (1991), *Natural Resources of Sri Lanka: Conditions and Trends*, Colombo: Natural Resources, Energy and Science Authority of Sri Lanka

Banaszak, S., U. Chakravorty and P. Ljeung (1999), 'Demand for ground transportation fuel and pricing policy in Asian Tigers: a comparative study of Korea and Taiwan', *Energy Journal*, **20** (2), pp.145-65.

Barnes, D.F, A. Domdom, V. Peskin, and H. Peskin (2002), 'Rural electrification and development in the Philippines: measuring the social and economic benefits'. ESMAP Report 255/02, World Bank, Washington, DC.

BCHydro (1995), 'Electricity Plan', Vancouver, British Columbia.

Black and Veatch (1988), 'Thermal Generation Options', Report to the Ceylon Electricity Board, Colombo.

Bose, R.K., and V. Srinivasachary (1997), 'Policies to reduce energy use and environmental emissions in Delhi', *Energy Policy*, **23** (3), pp.1137-50.

CEB (Ceylon Electricity Board) (1992), Wind energy resources assessment for the Southern Lowlands of Sri Lanka, Colombo.

CEB (Ceylon Electricity Board) (1998a), 'Load forecast for 1998 generation expansion planning studies', Colombo.

CEB (Ceylon Electricity Board) (1998b), 'Report on long term generation expansion planning studies, 1999-2013'. Generation Planning Branch, Colombo.

CEB (Ceylon Electricity Board) (1999) 'Report on long term generation expansion planning studies, 1999-2013'. Generation Planning Branch, Colombo.

CEETA (1998), 'Implementation in Portugal of the ExternE accounting framework', Lisbon.

Central Bank of Sri Lanka (2003), *Annual Report*, Central Bank of Sri Lanka, Colombo.

Chandrasiri, S., (1999), 'Controlling automotive air pollution: the case of Colombo City', Economy and Environment Programme for Southeast Asia, Research Report Series.

Chandrasiri, S. and S. Jayasinghe (1998), 'Health effects of vehicular emissions in Colombo', UC-ISS Project Working Paper Series, 9805.

Chin, A.T.H. (2000), 'Sustainable urban transportation: abatement and control of traffic congestion and vehicular emissions from land transportation in Singapore'. *Environmental Economics and Policy Studies*, **3** (4), pp. 355–80.

CONCAWE (Conservation of Clean Air and Water in Europe) (1989), 'Costs to reduce the sulphur content of diesel fuel', Report 10/89.

Connell Wagner Pty Ltd (2001), 'India coal port infrastructure study: coal importing economics'. Report to the Australian Department of Industry, Science and Resources, Melbourne.

Consultancy and Professional Services, (1999a), 'Feasibility Study of the Proposed Village Hydro Project at Handunella-Atulauda'. Report to the World Bank, Colombo.

Consultancy and Professional Services, (1999b), 'Feasibility Study of the

Proposed Village Hydro Project at Pathavita'. Report to the World Bank Colombo.

Copeland, T. and V. Anitkarov (2001), *Real Options: A Practitioner's Guide*, New York: Texere.

Dahl, C. (1986), 'Gasoline demand survey', *Energy Journal*, **7** (1), pp. 67–82.

Dahl, C. (1994), 'A survey of oil product demand elasticities for developing countries', *OPEC Review*, 18 (1), pp. 47–66.

Dahl, C. and J. Sterner (1991), 'Analysing gasoline demand elasticities: a survey', *Energy Policy*, July, pp. 203–10.

de S. Indraratna, A.D.V. (1994), 'Economic and environmental implications of petrol/diesel pricing', University of Colombo.

de Silva, A.P.D.S. (1989), 'Impact of energy price changes on transport', in E.N. Wijeratne, (ed.), *Energy Pricing Options in Sri Lanka*, ILO.

Dheerasinghe, K.G.D.D. and D.S. Jayaweera (1997), 'Economic, social and environmental demands on transport: past performance and emerging challenges', in Sri Lanka Transport Sector Study, Volume II, South Asia I Infrastructure Division, World Bank Report 16269-CE, Washington, DC.

Diandas, J. (1988), 'Notes for benefit comparison of government sector and private sector supplied bus services', in *Aspects of Privatisation in Sri Lanka*, Friedrich Ebert Stiftung, Colombo.

Diandas, J. (1989), 'Bus managers and users vis-à-vis ownership, regulation, competition and systems', Proceedings, European transport and Planning, 17th Summer Annual Meeting on Transport in Developing Countries, University of Sussex, September 1989.

Durstewitz, M., C. Ensslin, B. Hahn and M. Hoppe-Kilpper (2001), 'Wind energy development', Institut für Solare Energieversorgungstechnik (ISET), Kassel.

ECA (Economic Consulting Associates) and Robert Vernstrom (2003), 'EVN bulk power, distribution margin, retail consumer tariff design and development of an independent creditors model', Interim Report to EVN, Hanoi.

Electrowatt Engineering (1996), 'Thermal options study', Zurich.

Electrowatt Engineering (1997), 'Detailed feasibility study of the 3 x 300 MW West Coast Coal Project', Colombo.

Energy Conservation Fund (1996), Sri Lanka Energy Balance, Colombo.

Eskeland, G.S. and J. Xie (1998), 'Acting globally while thinking locally: synergy explored in air quality programmes for transport in Mexico City and Santiago', World Bank.

Faiz, A., C. Weaver and M. Walsh (1996), 'Air pollution from motor vehicles: standards and technologies for controlling emissions',

World Bank, Washington DC.

Fernando, C. and M. Munasinghe (1998), 'A real options framework to assess environmental policy, programs and strategy for GHG mitigation', World Bank, Washington DC.

Fernando, S. (1998), 'An assessment of the small hydro potential in Sri Lanka', Intermediate Technology Development Group, Colombo.

Foster, V., and J.-P, Tre. 2000. 'Measuring the impact of energy interventions on the poor—An illustration from Guatemala,' paper presented at a World Bank–U.K. Government conference, Infrastructure for Development: Private Solutions and the Poor, May 31–June 2, London. <http://www.ppiaf.org/conference/section1-paper4.pdf>

Freeman, A.M. (2000), 'Valuation of environmental health damages in developing countries: some observations', EEPSEA Special Paper, Biannual Workshop of the Economy and Environment Programme of Southeast Asia, Bangkok.

Frontier Economics (2003), 'Cost benefit analysis for renewable energy in Croatia', Report to the World Bank, London.

Fujime, K. (2002), 'LNG market and price formation in East Asia', Institute of Energy Economics, Tokyo.

Garbacz, C. (1989), 'Gasoline, diesel and motorfuel demand in Taiwan', *Energy Journal*, **10** (2), p.153.

Garrad Hassan (1999), Understanding the Risks of Financing Wind Farms. Bristol, UK.

Glaister, S., D. Graham and E. Hoskins (1999), 'Transport and Health in London' report for the NHS Executive, London.

Gomez-Ibanez, J.A. (1996), Bus industry Issues and options, in World Bank Transport Sector Study, South Asia I Infrastructure Division, Report 16269-CE, Volume II, p.126, Washington, DC..

Greene, D. (1992), 'Fuel use and fuel economy: how big is the rebound effect?' *Energy Journal*, **13** (1), pp. 117–143.

Grubb, N. and N. Meyer (1992), 'Wind energy: resources, systems and regional strategies', in T. Johansson et al. (eds), *Renewable Energy: Sources for Fuels and Electricity*, Washington DC: Island Press.

Gunawardena, P. et al. (1999), Yunnan Dachaoshan Power Transmission Project, in Environment and Economics in Project Preparation: Ten Asian Cases, Asian Development Bank, Manila.

Gunaratne, L. (1991), 'Making the investment: innovative ideas in alternative energy policy financing and services: annex market study commissioned by National Development Bank', paper presented at the FINESSE Workshop, October. 1991, Kuala Lumpur, Malaysia.

Halcrow Fox Associates (1996), 'Colombo urban transport study, working paper 8: transport costs', December.

Harvey, G. and E. Deakin (1992), 'Transportation and air quality', U.S. Department of Transportation, Federal Highway Administration, Policy Discussion Series #5, Washington, DC.

Henscher, D.A. (1995), 'Value of travel time savings in personal and commercial automobile traffic', Conference on Benefit-cost analysis, University of California at Irvine, 6–8 July 1995.

Holland, M. and P. Watkiss (2001), 'Estimates of the marginal external costs of air pollution in Europe: Benefits Table Database:BeTa', EU DG Environment, Brussels.

Horsnell, P. (1997), *Oil in Asia: Markets, Trading, Refining and Deregulation*, Oxford: Oxford University Press for the Oxford Institute of Energy Studies.

IDEA (International Development and Energy Associates) (2000), 'Sri Lanka electric power technology assessment', Background Report, Washington, DC.

IFC (International Finance Corporation) (1998), 'Sri Lanka: review of thermal options'. Washington, DC.

Indraratna, A. D. (1994),'Economic and environmental implications of petrol/diesel pricing,' University of Colombo, November.

IRG (International Resources Group) (2002), 'Impacts assessment and lessons learned', Washington, DC.

JICA (Japan International Cooperation Agency) (1984), 'Feasibility study of the Colombo-Katunayake Expressway', Colombo.

Jayaweera, D.S. (1998), 'Energy and environment-transportation sector', Transport Policy Centre, Colombo.

Jayaweera, D.S. (1997), 'Is there significant travel congestion in Colombo', paper presented at the Annual Meeting of the Chartered institute of Transport, Colombo, October.

Jayaweera, D.S. (1999), 'Effective steps towards traffic calming in developing countries: A case study for Metropolitan Region of Colombo in Sri Lanka', Ph.D. Thesis, MIT.

Jayaweera, D.S. (2000), 'Transport pricing and charges in Sri Lanka', Transport Studies and Planning Centre, Colombo.

Jayewardene, P. and V. Perera (1991), 'Sri Lanka: the rural electrification problem', Report by Power and Sun (Pvt) Ltd, Colombo.

Jeepma and Munasinghe (1998), *Climate Change Policy*, London, UK: Cambridge University Press.

Kojima, M., C. Brandon and J. Shah (2000), 'Improving urban air quality in South Asia by reducing emissions from two-stroke engine vehicles', World Bank South Asia Environmental Unit, Washington, DC.

Kumarage, A.S. (1992), 'Estimation of operational vehicle fleet in Sri Lanka, for road user charges study', Transport Studies and Planning Centre, Colombo.

Kumarage A.S. (1993), 'Estimates of passenger car units', Proceedings, Annual Sessions, Institute of Engineers, Sri Lanka.

Kumarage, A.S. (1996), 'Traffic engineering and traffic management', *Journal of the Sri Lanka Energy Managers Association*, **7** (1),pp.15–18.

Kumarage, A.S. (1999), 'Assessment of Public Investments in the Transport Sector', Appendix A1, University of Moratuwa.

Kumarage, A. S. (1999a), 'Managing the demand for cars in cities', Proceedings, National Conference, "Cities for All", World Habitat Day 1999, Sri Lanka.

Kumarage, A.S. (2000), 'Review of household income and public transport', Report to the World Bank, Colombo.

Kumarage A.S. and W. Wijertane (1999), 'Estimates of operational vehicle fleet in Sri Lanka 1998 update', University of Moratuwa.

Kumarage A.S. et al. (2000), TRANSPLAN V2: A regional traffic estimation model', University of Moratuwa.

LIFE (Lanka International Forum on Sustainable Development) (1998), 'Linkages between Economic Policies and the Environment in Sri Lanka', Colombo.

Lofgren, A., and H. Hammar (2000), 'The phase-out of leaded gasoline in the EU: a successful failure?', *Transportation Research*, **D**, 419–31.

Lovei, M. (1997), 'Phasing out Lead from Gasoline in Central and Eastern Europe', World Bank, Washington DC.

Lovei, M. (1998), 'Phasing out lead from gasoline: worldwide experiences and policy implications', World Bank Technical Paper 397, Washington DC.

Lutkepohl, H. (1991), Introduction to Multiple Time Series Analysis, New York: Springer-Verlag,.

Lvovsky, K., G. Hughes, D. Maddison, B. Ostrp and D. Pearce (2000), 'Environmental costs of fossil fuels, A rapid assessment method with application to six cities', World Bank, Environment Department, Paper 78, Washington, DC.

Masse, R. (2001), 'Impact on Poverty and Gender of Rural Electrification Programs in Sri Lanka', Colombo.

Matsui, K. (1998), 'Global Demand Growth of Power Generation, Input Choices and Supply Security', *Energy Journal*, **19** (2), p.93.

McRae, R. (1994), 'Gasoline demand in developing Asian countries', *Energy Journal*, **15** (1),P.143.

Meier, P. (1997), 'Economic Analysis of the Haryana Power Sector

Restructuring and Reform Program', World Bank, New Delhi.

Meier, P. (1999), 'India: Environmental Issues in the Power Sector: A Case Study of Haryana', World Bank, Washington, DC.

Meier, P. (1999), 'Economic Analysis of the Andhra Pradesh Power Sector Restructuring Program', World Bank, New Delhi.

Meier, P. (2003a), 'Economic and Financial Analysis of Community-based Small Hydro Projects, Vietnam System Efficiency Improvement, Equitization and Renewables Project', World Bank, Washington, DC.

Meier, P. (2003b), 'Economic Analysis of Solar Home Systems: A Case Study for the Philippines', World Bank, Washington, DC.

Meier, P. (2003c), 'Sri Lanka Energy Service Delivery Project: Economic analysis for the Implementation Completion Report', May, World Bank, Washington, DC.

Meier, P. (2004), 'Economic Analysis of the China Renewable Energy Scaleup Programme (CRESP)', World Bank, Washington, DC.

Meier, P. and M. Munasinghe (1994), Incorporating Environmental Concerns into Power Sector Decision-making: A Case Study of Sri Lanka', World Bank Environment Department Paper #6, Washington, DC.

Meier, P. and M. Munasinghe (1995), 'GHG Mitigation Options in Sri Lanka', *Energy Journal,* **16** (4) pp. 79–108.

Meier, P., and M. Munasinghe (2003a), 'Sri Lanka Electric Power Technology Assessment (SLEPTA)', ESMAP Report 262-03 (Vol III), Washington DC (www.esmap.org).

Meier, P., and M. Munasinghe (2003b), 'Sustainable transport options for Sri Lanka'. ESMAP Report 262-03 (Vol I), Washington, DC. (www.esmap.org).

Meier, P., and M. Munasinghe (2003c), 'Global Overlay of GHG Mitigation Options in the Sri Lanka Power Sector', ESMAP Report 262-03 (vol II), Washington DC. (www.esmap.org).

Meier, P., M. Munasinghe, and T. Siyambalapitiya (1997), 'Energy sector policy and the environment: a case study of Sri Lanka', in W. Cruz et al. (eds), *Greening of Economic Policy Reform*, Vol II, *Case Studies*, World Bank, Washington, DC.

MEIP (Metropolitan Environmental Improvement Programme) (1993), 'Air quality management in Sri Lanka', Colombo, Sri Lanka.

MEIP (Metropolitan Environmental Improvement Programme), (1994), 'Clean Air 2000, action program', Colombo, Sri Lanka.

Milligan, M. (2003), 'Measuring wind plant capacity value', National Renewable Energy Laboratory.

Milligan, M. and B. Parsons (1997), 'A comparison and case study of

capacity credit algorithms for intermittent generators', National Renewable Energy Laboratory, NREL/CP-440-22591.
Ministry of Transport (2001), 'Fuel efficiency survey' Colombo.
Mitchell, G. K. (1997), 'Middle east fuel supply and gas exports for independent power generation'. Power Economics: Fuel Supply Forum.
Munasinghe, M. (1987), *Rural Electrification for Development*', Boulder, CO and London: Westview Press.
Munasinghe, M. (1992), 'Environmental economics and sustainable development, Paper presented at the UN Earth Summit, Rio de Janeiro, Brazil, and reproduced as Environment Paper No. 3, World Bank, Washington, DC.
Munasinghe, M. (2001), 'Sustainable development and climate change: applying the sustainomics transdisciplinary meta-framework', *International Journal of Global Environmental Issues*, Vol. **5** (1), pp. 13–55.
Munasinghe, M., *et al.* (1998), 'Linkages between economic policies and the environment in Sri Lanka', Working Paper Series, No.1, Lanka International Forum on Environment and Sustainable Development, Colombo, Sri Lanka.
Munasinghe, M. (2004), *Evaluation of the UNEP Dam and Development Project*, United Nations Development Programme, Nairobi.
Munasinghe, M., and J. Warford (1980), *Electricity Pricing: Theory and Case Studies*, Baltimore, MD: Johns Hopkins Press for the World Bank.
Munasinghe, M, and Rob Swart (2005), 'Primer on Climate change and Sustainable Development - facts, policy analysis and applications', Cambridge, UK: Cambridge University Press.
Myers, R., and S. Thompson (1993), 'Managing oil import price risk in Costa Rica: strategies and benefits', in S. Claessens and R. Duncan (eds), *Managing Commodity Price Risk in Developing Countries*, Baltimore, MD: Johns Hopkins Press for the World Bank.
Newberry, D. (1990), 'Pricing and congestion: economic principles relevant to pricing roads', *Oxford Review of Economic Policy*, **6** (2)
Newman, P., and J. Kenworthy (1990), *Cities and their Traffic*, London: Gower Publications.
Nexant (2002), 'Viability of Developing a Transmission System Interconnection between India and Sri Lanka', Report to USAID, South Asia Regional Initiative for Energy.
Nexant (2003), 'Sri Lanka Natural Gas Options Study,' Report to USAID, South Asia Regional Initiative for Energy.
Nexant (2004), 'Sri Lanka Energy Sector Master Plan,' Interim Report,

ADB TA 4113-SRI, prepared for the Asian Development Bank, April.

Nordhaus, W. (1991), 'The cost of slowing climate change: A survey', *Energy Journal*, **12** (1), pp. 50–65

Pace University, 1990. Environmental Costs of Electricity, Centre for Environmental Legal Studies, New York: Oceana Publications.

Papathanasiou, D. and D. Anderson (2001), 'Uncertainties in responding to climate change: on the economic value of technology policies for reducing costs and creating options', *Energy Journal*, **22** (3), pp.79–114.

Pigou, A. (1924), *The Economics of Welfare*, London: MacMillan.

Posch and Partners (1994), 'Sri Lanka micro hydro feasibility study', Report to the Asia Alternative Energy Unit, World Bank, Washington DC.

Ranasinghe, A.P. (1989), 'Impact of fuel price changes on Transport', in E.N. Wijeratne, (ed.), *Energy pricing Options in Sri Lanka*, ILO.

Ratnasiri, J. (ed.) (1998), 'Final Report of the Sri Lanka climate change country study', Ministry of Forest and Environment, Environment Division, Colombo.

RDA (Road Development Authority) (1997), 'Colombo Katunayake expressway', Environmental Impact Assessment Report, Colombo.

Reyes, Carmela. 2001. 'Twin Girls Killed in Bulacan Fire', Inquirer News Service, December 23. <www.inq7.net/met/2001/dec/23/met_6-1.htm>

Rowe, R.D. et al. (1995), 'The New York electricity externality study', Dobbs Ferry, NY: Oceana Publications.

Salsena, S. and V. Dayal (1998), 'Valuing air pollution in India', *Pacific and Asian Journal of Energy*, **8** (2), 207–20.

Samarakoddy, R.P. (1999), 'Air quality monitoring in Sri Lanka', Proceedings, Asian Aerosol Data Synthesis Workshop, 13-15 December, New Delhi

Samarakoddy, R.P. (2001), 'Air pollution in Colombo', National Building Research Organisation, Colombo.

Samaraweera, C.P. (1989), 'Determinants of petroleum demand in Sri Lanka', in E.N. Wijeratne, (ed.), *Energy Pricing Options in Sri Lanka*, ILO.

SCADA (Sone Command Area Development Authority) (1999), 'India: global overlay for Bihar', Report to the World Bank, February.

Schafer, A. and D. Victor (1999), Global passenger travel: implications for carbon dioxide', *Energy*, **24**, p.657–79.

Schipper, L. (1999), 'Flexing the link: transport sector policy and GHG emissions', World Bank, Washington DC.

Schwartz, J. (1994), 'Environmental benefits of phasing out lead in gasoline', *Environmental Research*, **66**, 105–24.

Senanayake, M.P., Samarakoddy, R.P. et al. (1999), 'Is Colombo choking its children: A Case-control analysis of hospital attendance and air pollution', National Building Research Organisation and Faculty of Medicine, University of Colombo.

Shoup, D. and S. Stark (2000), 'The parking of nations', Access Almanac # 17, University of California, Berkeley.

Shresta, R., W.J.L.S. Fernando and Rabin Shrestha (1998), 'Environmental emission mitigation potential of efficient electrical appliances in Sri Lanka', *International Journal of Energy Research*, **22**, 923–33.

Sivagnasothy, V. and D. McCauley (2000), 'Environmental considerations in the environmental impact analysis of the Katunayake Expressway in Sri Lanka', in J. Rietbergen-McCracken and H. Abaza, (eds), *Environmental Valuation: A Worldwide Compendium of Case Studies*, London: Earthscan Publications.

Smith, K. R. (1993), 'Fuel combustion, air pollution exposure and health: the situation in developing countries', *Annual Review of Energy and Environment*, **18**, 529–66.

Sone Command Area Development Authority (SCADA) (1999), 'India, global overlay for Bihar', Report to the World Bank, New Delhi.

Spadaro, J. and A. Rabl (1998), 'Air pollution damage estimates: the cost per kg of pollutant', Ecole des Mines de Paris, Centre d'Energetique, Paris.

Sustainable Energy Advantage, LLC (2003), 'Using wind power to hedge volatile electricity prices for commercial and industrial consumers in New York', Report to the New York State Energy Research and Development Authority, Albany.

Toichi, T. (1994), 'LNG development at the turning point and policy issues for Japan'. *Energy Policy*, **22** (5) 371–77.

Thomas, V. (1995), 'The elimination of lead in gasoline', *Annual Review of Energy and Environment*, **20** 301–24.

Toh, R.S. (1996), 'Experimental measures to curb road congestion in Singapore: pricing and quotas', *Logistics and Transportation Review*, **28** (3) p.289–317.

Tokyo Electric Power Services Ltd (1998), 'Kerawalapitiya CCCT detailed feasibility study', Tokyo.

Transport Studies and Planning Centre (1996), 'Study of transportation services provided by school and office vans', Colombo.

UK Committee on the Medical Effects of Air Pollution, Department of Health (1998), 'Quantification of the effects of air pollution on health in the UK'.

UNEP (United Nations Environment Programme) (1998), 'Global opportunities for reducing the use of leaded gasoline'.

University of Moratuwa (2000), 'Traffic demand estimation for the proposed Colombo–Katunayake Expressway', Report to Board of Investment, Colombo.

USEPA (1985), 'Costs and benefits of reducing lead in gasoline; Final Regulatory impact estimate', Office of Policy Analysis, Washington, DC EPA-230-05-85-006.

USAID (1995), 'Integrated resource plan for Andhra Pradesh', Report to Andhra Pradesh State Electricity Board, Hyderabad.

Vivekaphirat, P. (2001), 'Experience of Thailand in improving fuel quality', Proceedings, Regional Workshop: Fuel Quality and Alternative Fuels, Delhi.

W. S. Atkins and University of Moratuwa (1995a), 'A masterplan for the development of transport in the Colombo metropolitan region', Colombo Urban Transport Study, Working Paper 25.

W. S. Atkins and University of Moratuwa (1995b) 'Feasibility study of an inland container depot', Colombo Urban Transport Study, Working Paper 12.

W. S. Atkins and University of Moratuwa (1999), 'Colombo Urban Transport Study, Stage 2', Ministry of Transport & Highways, Colombo.

Walpita, N. (1997), 'Biomass development and environmental consequences', Colombo.

Walsh, M. (1992), 'Motor vehicle pollution control in Sri Lanka, a preliminary strategy for progress in Colombo', Report to the Natural Resource and Environmental Policy Project, Colombo, July.

Wijewardene, R., and P.G. Joseph (1999), 'Growing our own energy: complementing hydro-power for sustainable thermal energy and rural unemployment in Sri Lanka', Colombo.

Wilbur Smith Associates (1998), 'Southern transport corridor project', December.

World Bank (1992), 'Mexico: transport air quality management in the Mexico City area', 10045-ME, Washington, DC.

World Bank (1994), 'Thailand: mitigating pollution and congestion impacts in a high-growth economy', Report 11770-TH, Washington, DC.

World Bank (1995), *World Tables 1995*, Washington DC : World Bank.

World Bank (1996a), 'Petroleum Sector Study', Energy and Project Finance Division, Country Department I, South Asia Region, Washington, DC.

World Bank (1996b), 'Sri Lanka, transport sector study', South Asia I Infrastructure Division, Report 16269-CE, Washington, DC.

World Bank (1997a), 'Project appraisal document', Energy Services Delivery Project, Report 16063-CE, Energy and Project Finance Division, Country Department 1, South Asia Region.

World Bank (1997b), 'Economic analysis of the Haryana power sector reform and restructuring programme', New Delhi.

World Bank (1997c), 'China 2000: Clear water, blue skies', Washington, DC

World Bank (1997d), 'Guidelines for climate change global overlays', Global Environment Division, World Bank Environment Department Paper 47, Climate Change Series, Washington, DC.

World Bank (1997e), 'Economic analysis of the Andhra Pradesh power sector reform and restructuring programme', New Delhi.

World Bank (2000), 'Comments on the proposed fuel quality and vehicle emission standards', Washington, DC.

World Bank (2001),'The California power crisis: lessons for developing countries', Washington, DC <rru.worldbank.org/PapersLinks/
Open.aspx?id=3549>

World Bank (2004a), 'Environmental issues in the power sector: long-term impacts and policy options for Rajasthan', Washington, DC (www.esmap.org)..

World Bank (2004b), 'Environmental issues in the power sector: long-term impacts and policy options for Karnataka', Washington DC (www.esmap.org).

World Bank (2005), 'Yemen: Household energy supply and use', World Bank, Washington, DC.

Yangsheng, Z. (2001), 'Improving fuel quality in China'; Proceedings, ASEAN Regional Workshop on Fuel Quality and Alternative fuels, Delhi, May.

Zaidi, S. M. (2001), 'Country experience of Pakistan in improving fuel quality', Proceedings, ASEAN Regional Workshop on Fuel Quality and Alternative Fuels, Delhi, May.

Index